NEUROTECHNOLOGY AND THE END OF FINITUDE

CARY WOLFE, SERIES EDITOR

45 Neurotechnology and the End of Finitude
Michael Haworth

44 Life: A Modern Invention
Davide Tarizzo

43 Bioaesthetics: Making Sense of Life in Science and the Arts
Carsten Strathausen

42 Creaturely Love: How Desire Makes Us More and Less Than Human
Dominic Pettman

41 Matters of Care: Speculative Ethics in More Than Human Worlds
Maria Puig de la Bellacasa

40 Of Sheep, Oranges, and Yeast: A Multispecies Impression
Julian Yates

39 Fuel: A Speculative Dictionary
Karen Pinkus

38 What Would Animals Say If We Asked the Right Questions?
Vinciane Despret

37 Manifestly Haraway
Donna J. Haraway

36 Neofinalism
Raymond Ruyer

35 Inanimation: Theories of Inorganic Life
David Wills

34 All Thoughts Are Equal: Laruelle and Nonhuman Philosophy
John Ó Maoilearca

33 Necromedia
Marcel O'Gorman

32 The Intellective Space: Thinking beyond Cognition
Laurent Dubreuil

31 Laruelle: Against the Digital
Alexander R. Galloway

30 The Universe of Things: On Speculative Realism
Steven Shaviro

29 Neocybernetics and Narrative
Bruce Clarke

28 Cinders
Jacques Derrida

27 Hyperobjects: Philosophy and Ecology after the End of the World
Timothy Morton

26 Humanesis: Sound and Technological Posthumanism
David Cecchetto

25 Artist Animal
Steve Baker

24 Without Offending Humans: A Critique of Animal Rights
Élisabeth de Fontenay

(continued on page 200)

NEUROTECHNOLOGY AND THE END OF FINITUDE

MICHAEL HAWORTH

posthumanities 45

University of Minnesota Press
Minneapolis
London

Portions of chapter 3 were previously published as "Synchronicity and Correlationism: Carl Jung as Speculative Realist," *Speculations: A Journal of Speculative Realism* 3 (2012): 189–209. Portions of chapter 4 were previously published as "Telepathy and Intersubjectivity in Derrida, Husserl, and Levinas," *Journal of the British Society for Phenomenology* 45, no. 3 (2015): 253–66. Reprinted by permission of Taylor & Francis Ltd, www.tandfonline.com.

Published by the University of Minnesota Press
111 Third Avenue South, Suite 290
Minneapolis, MN 55401-2520
http://www.upress.umn.edu

ISBN 978-1-5179-0331-2 (hc)
ISBN 978-1-5179-0332-9 (pb)

A Cataloging-in-Publication record for this book is available from the Library of Congress.

Printed in the United States of America on acid-free paper

UMP BmB 2018

FOR HELEN AND BERNARD

Rapidly we approach the final stage of the extensions of man.

MARSHALL MCLUHAN, *UNDERSTANDING MEDIA*

To eradicate the gap, to put an end to the scandal of the interval of space and time that used to separate man so unacceptably from his objective: all this is well on the way to being achieved. But at what cost?

PAUL VIRILIO, *OPEN SKY*

CONTENTS

Acknowledgments *xi*

Introduction 1

1 The Idea Becomes a Machine That Makes the Art 13

2 Intellectual Intuition and Finite Creativity 45

3 *Unus Mundus* 87

4 Techno-Telepathy and the Otherness of the Other 127

Notes *169*

Index *193*

ACKNOWLEDGMENTS

This book would not have been possible without the teaching and friendship of Alex Düttmann. I could not begin to map out the extent of his influence on my thinking but it can be felt on every page. The second major influence on the writing of this book is Chris Haworth, who has been good enough to read several draft versions of each chapter, and each time made improvements with penetrating comments and important suggestions. I owe my awareness of the field of brain–computer music interfacing to Chris, as well as a great deal else, so his influence on the genesis and development of this project cannot be overstated.

Others who have read the manuscript, either in whole or in part, at various stages and contributed significantly to its development are Alberto Toscano, Simon O'Sullivan, Martin McQuillan, Howard Caygill, and the anonymous readers for the University of Minnesota Press and the journals in which sections have been published previously. Simon deserves special mention for having been particularly influential in how I approached my reading of Jung, as does Alberto for his precisely targeted criticisms, which I hope to have responded to adequately. I also express my thanks to Cary Wolfe for his time and generosity. His enthusiasm for the project has been of enormous encouragement. Likewise, my thanks to Doug Armato for all of his time and support in bringing the book to fruition.

My parents, Anne and Bill Haworth, have supported me in immeasurable ways throughout the researching and the writing of this book, and I want to convey my profound thanks to them. Everything I have done I owe to their influence. Finally, this book has been a long time in the making and without the love, support, and friendship of my wife Helen I could neither have started nor finished it.

INTRODUCTION

It has become increasingly common in recent years to hear predictions, whether fearful or celebratory, about the alarmingly transformational future in store for human subjectivity. Given the unprecedented advances made over the last few decades in genetics, neuroscience, and information technology, as well as anticipated innovations in nanotechnology and robotics, no aspect of the human condition seems impervious to radical alteration. Optimistic observers see the rate at which previously insurmountable bodily deficiencies and frailties are now routinely overcome, and new powers or capabilities granted, to be evidence that there is no inherent biological limitation that cannot in principle be transcended. Such eschatological discourses welcome the potential for new technologies to transform and improve our inherited situation by relieving suffering and reversing aging, as well as boundlessly enhancing our intellectual capabilities through chemical and technological interventions into the brain. As the famous inventor and seer of the digital age Ray Kurzweil claims, "[we're] going to gradually merge and enhance ourselves. In my view, that's the nature of being human—we transcend our limitations."[1] In progressing toward "this future man, whom the scientists tell us they will produce," as Hannah Arendt so presciently wrote sixty years ago, humankind seeks to exchange "human existence as it has been given, a free gift from nowhere (secularly speaking) . . . for something he has made himself."[2] By overcoming facticity with a state of pure self-creation we would finally have surpassed our finitude.

The foundations for this increasingly intimate integration of the biological and the technological were laid by the cybernetics movement of the 1940s and 1950s, and brought into popular consciousness by contemporary science fiction writers. However, the vast leaps made in our scientific understanding of the brain and the human genome, along with advances in computing, have led to the recent acceleration of this tendency. An exemplary case is the emergence of quasi-miraculous neurotechnologies, which establish a direct channel of communication between a brain and a machine. From a philosophical perspective, their novelty and significance lies in the fact that they appear to present the possibility of overcoming the most general limitation of human subjectivity, namely, our confinement to the interiority and privacy of the mind. Such technologies represent a passage to the limit at which subjective interiority meets that which is irreducibly external to it. Rather than prosthetically extending particular faculties or organs and triumphing over specific finite constraints, they act on the very limitation of finite selfhood as such. Our restriction to a self-enclosed interiority, known as ipseity in the philosophical literature, is, as F. W. J. Schelling says, the "original limitation" from which all experience derives.[3] It is the grounding and most general mark of our finitude, underlying and conditioning all particular limitations, whether bodily or intellectual. Neurotechnologies therefore represent something of a terminal point in the narrative of human enhancement and a testing ground for speculative enquiries into the extent of our abilities to technologically transcend our limitations, which humanists such as Kurzweil consider to be of the very essence of human nature.[4]

By allowing for mental events to automatically bring about physical events, or for the thought of an object to be automatically externalized, the technologies in question intervene at the very limit point between interiority and exteriority, offering a truly unprecedented immediacy: an intention automatically brings about its own realization, a desire its own fulfillment, an idea its own expression. In short, possibility coincides with actuality with no delay and no effort expended. The primary question for us will be whether this amounts to a surmounting of the finite distance separating that which belongs to the mind and that which exists outside of it and beyond its power (the *I* and the *not-I* in the language of Fichte). What remains of the interior-exterior structure of finite experience once we are endowed with a faculty of intervening directly with the mind into the outside world or other minds? How does this affect human creativity and how will it alter our sense of communicating with others?

Before proceeding any further it will be necessary to specify in detail exactly what is meant by this term *neurotechnologies* and in what respects they represent genuine novelty. The name neurotechnology refers to any technology that establishes a direct communication pathway between the activity of the brain and an external device.[5] This channel can operate in both directions: "inwards" and "outwards." An example of the former is direct neural stimulation technologies, which transmit electrical or magnetic pulses to specific areas of the brain in order to treat the symptoms of neurodegenerative disorders. Another example is increasingly sophisticated neuroimaging techniques such as magnetic resonance imaging (MRI), which allow researchers and clinicians to map out the overall functional architecture of the brain and the neural correlates of higher order mental functions. The key instances of the latter "outward"-directed neurotechnologies are brain–computer interfaces (BCIs), which translate electrical brain signals into instructions for operating a computer-controlled device. It is with the latter type of technology, which allows for a direct psychical intervention into reality, that we will be particularly concerned throughout this book. Then, in the final chapter we will turn our attention to technologies that combine the two functions to allow for direct communication between one brain and another.

Brain–Computer Interfaces

With the operation of a standard technological instrument a whole chain of complex detours is involved in the transmission of the brain activity associated with the user's intention to the apparatus itself. The neural signals engage the peripheral nervous system, resulting in muscle contractions, which in turn bring about the desired action from the device. A brain–computer interface (BCI), on the other hand, sets up a *direct* communication channel between the brain signals and the tool, allowing the brain to bypass its regular organic pathways and send control information to an external apparatus, such as computer software or a prosthetic device, solely through its own activity and without the need for physical movement. A BCI therefore facilitates the psychical control of an electronic device by the operator explicitly *intending* the task to be performed. It works by recording brain activity, most commonly using an EEG machine, and uses signal processing and pattern recognition to translate this activity into a specific command or set of commands for the machine to execute.

The principle of harnessing brain activity in order to exercise remote control over a machine goes back at least to the mid-1960s, with physicist Edmond

Dewan's experiments with EEG at the Air Force Cambridge Research Laboratory in Bedford, Massachusetts. As Douglas Kahn describes in his book *Earth Sound Earth Signal,* Dewan devised a control mechanism for turning a standard desk lamp on and off by modulating his brainwaves while wearing electrodes on the surface of the scalp.[6] The system was powered by alpha waves, which is a particular frequency of brainwave detectable by EEG machine when the subject is in a relaxed state with eyes closed. The alpha waves were directed to electronic amplifiers, which strengthened the signal to the point where it could be used to trigger a switching mechanism, sending current to the lamp when alpha waves were present and cutting off when they were not. So the lamp was operated by the subject doing nothing more than drifting in and out of a state of relaxation, which, with practice, Dewan was able to employ in order to communicate in Morse code.

At almost exactly the same time as Dewan was conducting his experiments, neurophysiologist W. Grey Walter developed a procedure often considered to be the earliest example of a BCI, in which a slide projector was operated by brain activity.[7] In this case, rather than wearing an external electrode cap as Dewan had done, the electrodes were connected directly to the patient's motor cortex while undergoing unrelated brain surgery. The patient was given the task of pressing the button to progress through the slides on a standard slide projector while Walter recorded the brain activity associated with the hand movement. In the next phase, the electrical signals from the brain were channeled directly to the slide mechanism so that the projector advanced through the slides whenever the brain activity corresponding to the intention to press the button was detected.

However, despite these early innovations it is computer scientist Jacques Vidal's laboratory at UCLA that is generally considered to be the birthplace of the brain-computer interface (a term Vidal coined in the 1970s). Where Dewan's method used continuous signals from the brain to control an external device and Walter's used motor signals from invasive implants, Vidal's innovation was to make use of evoked potentials (also known as event-related potentials, or ERPs) as control information.[8] These are short, distinctive spikes of electrical activity observable in the EEG readings occurring at particular points following a sensory stimulus, such as a visual or auditory message or a tap on the arm. What the discovery of evoked potentials allowed the BCI researchers to do was to identify the points in the EEG patterns when the user is focusing on a particular sensory stimulus, commonly an image chosen from a selection on a screen. The BCI system interprets and classifies the brain activity corresponding to this choice and matches it to a particular command, which is then executed by the machine. The advantage of this method over Walter's earlier method is that it is noninvasive and therefore safer and more practical, while its advantage over Dewan's procedure is that rather than a simple on/off instruction it allows for multiple options, thus enabling much more accurate control over more specific tasks. Variants of Vidal's method are still widely used in BCI research today, where the output is controlled by the user exercising selective attention, an example being the control of a computer cursor or robotic device by focusing on a series of flashing symbols corresponding to up, down, left, right, and select.[9] Here the output is entirely controlled by the user's cognitive choice, which *of itself* brings about its own realization. In addition to assistive technology applications such as wheelchairs and prosthetics this method has also been used as the basis for a communication vehicle for patients with locked-in syndrome.[10]

While research in BCI technology has been primarily directed toward clinical applications for disabled users who have lost most or all motor functions, technology developers are already beginning to find commercial uses. There are now several consumer BCIs available to buy relatively cheaply, the earliest example being the MindWave Mobile headset which was launched in 2010 by the consumer electronics company NeuroSky. It looks similar to the head microphones worn in call centers or live music performances but instead of using sound as the input source it interprets brain signals, detecting and processing brainwaves associated with attentive and meditative states and using this data for various applications. Closer to Edmond Dewan's brainwave control system than the more advanced BCIs used in clinical research, they are primarily used as a training tool to aid concentration or as an input device for video games, such as the amusingly titled "Throw Trucks with your Mind," where calm, focused attention "allows you to pick up and hurl objects at opponents, turn invisible, put up a force field, or jump extremely high."[11] In fact the emergent field of Neurogaming, as it has inevitably become known, even has its own annual Silicon Valley conference and expo, begun in 2013.

Alongside selective attention tasks such as those previously outlined, the other most common mental strategy used to generate control information for a BCI system is motor imagery. This is when the subject visualizes the movement of a limb, which produces a recognizable pattern of brain activity close to that associated with the actual limb movement itself. The information is interpreted by the BCI and used to send a command to a machine, for example "move left" by imagining left hand movement. In a remarkable and well-publicized example of this procedure, at the University of Pittsburgh in 2007, laboratory monkeys learned to operate a robotic arm with the brain alone in order to feed themselves.[12] With microelectrode arrays implanted directly into their motor cortices, the monkeys were placed in a restraining harness to prevent them from using their own arms while a food source was placed in front of them. The motor signals in the brain associated with arm movement were directed to the machine arm, just as in Walter's early experiment with the slide projector, and over time they learned to manipulate it with an astonishing degree of fluency, as a remarkable YouTube video of the experiment shows.[13]

Invasive methods such as these inevitably yield more successful results than noninvasive methods using EEG or similar owing to the proximity of the electrodes to the brain activity. However, the applicability of invasive BCI methods to human users is limited for obvious reasons of safety and practicality. For this reason many of the participants who have taken part in invasive BCI research have been patients already undergoing brain surgery for some preexisting condition or brain injury. In such cases, during surgery while the skull is opened and the brain exposed, an electrode array called an electrocortigram (ECoG) is placed directly onto the surface of the cerebral cortex. Given the risks involved, the current research streams working with ECoG are exclusively geared toward helping individuals with paralysis regain some form of engagement with their environment. However, researchers such as Daniel Moran at Washington University in St. Louis believe that as the technology and surgical procedures are perfected, a version of this technology will inevitably find its way to the marketplace. Moran, who has been involved in developing invasive BCI systems in humans and primates for several years, predicts, with casual brevity, that "[eventually] we'll have a little piece of Saran wrap with telemetry. We'll drill a small hole in the skull, pop the bone out, drop the device in, replace the bone, sew up the scalp and you'll have what amounts to Bluetooth in your head that translates your thoughts into actions."[14]

Cognitive Imaging

As well as translating brainwaves into desired actions and bringing about the immediate execution of a command associated with the user's will, it is also possible, to a certain extent, to analyze a person's EEG pattern to decode what object he or she is looking at. The activity in the brain evoked by observing particular types of objects is relatively consistent, so the EEG signals recorded when concentrating on an image, say of a horse, will be the same when looking at another image of a horse. This information can then be stored and used as a point of reference and from this an extensive database or dictionary has been generated matching patterns of EEG curves to objects. Therefore, not only can it be accurately determined by EEG signal processing *when* the subject is training his or her attention on something, as in the BCI systems described above, it can also be approximately predicted *what* the subject is focusing on. It is entirely conceivable that such information may eventually be utilized in a future BCI, where the computer would learn to recognize the patterns of activity corresponding to certain choices *directly*, without the need for the symbolic representations on a screen.

Of course, these are relatively crude approximations corresponding to certain general categories of object. What the EEG machines certainly cannot do is to reproduce an individual visual stimulus or mental picture based solely on brain activity. However, a different strand of neurotechnology research is working on developing precisely this type of reconstruction. In an extraordinary series of experiments at the University of California, Berkeley, neuroscientist Jack Gallant and his team have been able to digitally reconstruct live, dynamic visual experiences from the brain using functional magnetic resonance imaging (fMRI) combined with computational models.[15] The participants watch short film clips while lying flat for several hours in an MRI scanner, which maps changes in blood flow in the brain associated with neural activity. Gallant's team has developed a sophisticated computer algorithm which analyzes the film clips and the MRI patterns and searches for correlations between the configuration of shapes, colors, and movements in the moving images and the concomitant patterns of activity in the subjects' visual cortex. Over time the computer learned to recognize matches between image patterns and brain activity and used this to predict how a given sequence of images would affect the visual cortex of the viewer. In the second phase of the experiment the participant's brain activity was measured while watching a new set of films, and the computer program was able to decode the brain activity and generate a reconstruction of the visual stimulus. The result is that the computer displays on screen a highly distorted but remarkably accurate approximation of what the subjects were seeing. The reconstruction is based on a database of movie clips, which the computer uses as raw materials, collaging fragments together that have been identified as close matches to the film being viewed as identified from the neural activity. I will be using the term *cognitive imaging* to refer to this process that is aimed at visualizing cognitive states.

The ethical and political concerns that emerge with the rise of such an unprecedented regime of technological intrusion are as obvious as they are enormous. We could have a field day predicting the uses advertisers may make of neurotechnologies in some bleak dystopian future where capital has gained access to our intimate thoughts, or prophesying Philip K. Dick-esque state-sponsored neuromonitoring. Already, disturbing reports have emerged that the U.S. military is actively exploring the use of a brain scanning device designed by "applied neuroscience" technology developers Veritas Scientific to discover insider threats among their Iraqi and Afghan colleagues.[16] The technology utilizes what is known in neuroscience as the P300 response, which is an event-related potential (ERP) component associated with image recognition. As described above, ERPs are fluctuations in the brain signal detectable in EEG readings that occur in response to an external stimulus, and the P300 wave is evoked in response to an emotionally significant sensory input, such as something unexpected or recognizable. The technology, menacingly named HandShake, works by presenting the person under interrogation with a series of images, mostly benign, but some of which are designed to catch out enemy operatives: faces of known terrorists, diagrams showing how to build an improvised explosive device (IED), and so on. If the subject's brain shows signs of recognition when confronted with certain loaded images this is considered strong grounds for suspicion. As the reaction in the brain occurs before the subjects are consciously aware of what they have seen, they are not able to modulate their response to cheat the system. Of course, it does not take a paranoiac to infer that such "enhanced security measures" could be on their way to major airports at some point in the near future.

Although attending in detail to the explicitly political ramifications of neurotechnology is undoubtedly a task of considerable importance, this is not my primary concern here. Discussions surrounding the relation between politics and the brain have flourished in the last decade. The University of Edinburgh recently founded a Neuropolitics Research Lab dedicated to interdisciplinary analysis of the role of the brain in social behaviors and political attitudes, and there has been important work by Catherine Malabou, Bernard Stiegler, and others on the relationship between the brain and neoliberal capitalism. Such discussions, while not taking center stage in this book, hover in the margins at several key junctures.

Thinking the Gap

My primary interest in neurotechnologies, and the reason why their arrival is of such singular theoretical significance, is that rather than acting as a supplement or corrective to overcome a particular, localized deficiency, they intervene at the very limit between subjective interiority and that which exists independently of it. This is nothing less than the limit of our original finitude and if, as thinkers from Freud and Marshall McLuhan to Bernard Stiegler and Andy Clark have demonstrated, technology is essentially compensatory, forming an open-ended chain of prosthetic extensions to—always provisionally—counterweigh an originary finitude, what we are apparently confronted with at the advent of an apparatus that can harness and instrumentalize brain activity itself is the final term in the series: the technology to bring an end to technologies. According to Slavoj Žižek, the "direct short circuit between mind and reality" brought about by brain-computer interface technologies "implies the prospect of a radical closure."[17] This closure is nothing less than that of the finite interval that necessitates the technical tool in the first place. Žižek claims that this coincidence of thought with event essentially amounts to a form of spontaneous creative activity equivalent to what Kant called intellectual intuition. With the arrival of such a creative power the receptivity that characterizes human cognition, which Kant considered central to our finitude, would be surmounted.

However, what would be the consequences of such an overcoming and how is it to be conceived? After all, this gap between thought and that which is external to it is no mere impediment, rather, as we will see through extended readings of Kant, Heidegger, and Freud, thought only exists in the space opened by this gap. As Adorno writes, all thought is "mediated by the whole flow of conscious life in the knowing subject" and this mediation is not something we must seek to overcome in the aim of ever-greater objectivity, for that would ultimately amount to thought's dissolution.[18] The inescapability of one's own individual point of entry onto the world (or our original finitude), including "prejudices, opinions, innervations, self-corrections, presuppositions and exaggerations,"[19] that perennially keep us at a distance at one remove from an immediate grasp of "reality," is in fact the enabling source of insight and knowledge rather than its obscuring fog. A pure adequation of thought to object, or divine insight, would then be mere repetition. In a different register, Thomas Metzinger writes that "for beings like ourselves the fact that there is an irrevocable boundary between ourselves and our environment . . . is simply a fundamental feature of reality itself."[20] Our ability to firmly distinguish between what is "in me" and what is "not me" is one of the basic conditions of our experience, so the dissolution of this boundary could only be envisaged as a catastrophic loss of agency rather than an enhancement of it.[21]

The question of whether the gap between the "in here" of the mind and the "out there" of reality can be overcome, or thought otherwise, without dissolving one into the other will be the engine for everything that follows. The book as a whole thus resembles an extended thought experiment, testing the idea that we may ultimately surpass our finitude via neurotechnologies and hyperbolically envisioning the consequences of doing so. This idea is systematically pursued across four distinct but interrelated fields: first, art, by means of a close reading of the idealist aesthetics of Benedetto Croce, in which idea and expression are one and the same process; second, philosophy, through an exegesis of Kant's account of the infinite creativity of the Supreme Being; third, psychoanalysis, via the juxtaposition of Freud and Jung's respective models of the self-world relation; finally, telepathy, where the idea of an immediate communion between minds is pitted against various models of the intersubjective encounter. Each of the four chapters is to a large extent self-contained, bearing its own "narrative arc," as it were, and its own set of theoretical resources, but the scope of the problem is progressively expanded as the book unfolds. It is in the work of Carl Gustav Jung (chapter 3) and Jacques Derrida (chapter 4), on the apparently paranormal themes of synchronicity and telepathy, that we find the paradigms for rethinking the relationship between the mind and its outside as not being structured by a gap.

Chapter 1 is concerned with the striving for immediacy and purity in creative expression that characterized several strains of modernist art theory and practice. This ideal of immediacy can take two forms: either art as pure concept, expunged of any aesthetic, expressive interference, or art as pure, disengaged expression, free from conscious direction. Springing from opposite poles of the idea-expression dichotomy, each tries to uncouple itself from the other. Neurotechnologies, when creatively employed, could be seen to realize either or both of these goals. Cognitive imaging technologies would allow for an idea to be automatically birthed into existence, and BCI technologies would act as an immediate outlet for creative expression. This question is investigated primarily through an extended, critical reading of the aesthetic philosophy of Benedetto Croce, where the separation of idea and expression comes to be problematized. However, because Croce only overcomes this duality by reducing both processes to the interior of the artist's mind, he becomes entangled in internal contradictions where the physical medium is both essential and inessential, internal and external to the formation of the work. By reading Croce against more recent work in cognitive psychology and philosophy of mind I seek to retain Croce's central insight that an idea is already its own expression but without surrendering to an untenable idealism. As case studies, I examine recent musical and sculptural applications for brain-computer interface technologies.

Chapter 2 pieces together the scattered references Kant makes to the concept of intellectual intuition, in the *Critique of Pure Reason* and elsewhere, in order to develop a rigorous model of a nonfinite, nonreceptive mode of cognition. This is, according to Žižek, the ultimate corollary of neurotechnologies, which bring about the immediate union of cognition and object. However, for Kant, the finite, receptive character of our cognition is not a merely contingent shortcoming that can be compensated for by other means, rather, it is a transcendental necessity. On the occasions when Kant does hypothetically envisage the overcoming of finitude, it is invariably depicted as catastrophe, for as Kant grandly discourses, "the inscrutable wisdom through which we exist is not less worthy of veneration in what it has refused us than in what it has allotted us."[22] This question of human finitude and the possibility—or desirability—of surmounting our receptive faculty of intuition with a creative one is further explored along two avenues: first, Heidegger's early interpretation of Kant, and second, Quentin Meillassoux's unexpected resuscitation of the concept of intellectual intuition in his 2006 book *After Finitude*.

Chapter 3 is made up of two parts juxtaposing the contrasting accounts Freud and Jung present of the relation between psyche and world. The first part uses the terms of Freudian psychoanalysis to explore the consequences of the immediate, automatic fulfillment of wishes and realization of desires. Throughout his corpus Freud places great significance on a series of strict oppositional distinctions which all derive from the core opposition between internal and external. The ego only comes to be individuated through decisively severing itself and its own activities from external reality. By undermining this distinction, neurotechnologies would apparently threaten our whole psychical consistency. Jung on the other hand, erects no such firm distinction between internal and external. The second part of the chapter explores his work theorizing the archetypes of the collective unconscious, where Jung develops a neutral, monist ontology to which he gives the name *Unus Mundus*. I set this concept against that of intellectual intuition as a more useful and consistent model for rethinking, and thereby overcoming, the abyssal gap between mind and world.

In Chapter 4 the concept of telepathy is probed, prompted by the very recent emergence of brain-to-brain communication technologies. The question will be whether a techno-telepathic exchange would amount to a more direct form of communication that would enable an unprecedented degree of access to the other's interiority. Since the relation between *I* and other is of a profoundly different order to that between mind and world this last chapter serves to reframe and recast all of the questions and outcomes of the previous chapters. Through extended interpretations of some of Derrida's late texts the argument ultimately arrives at a conceptualization of telepathy that is not reducible to the communication of a message between an active sender and a passive recipient. The stable identity of *I* and other is undermined but without diminishing the distance between them. In this, as in each of the preceding chapters, the idea that new technological advances will allow us to break through some ultimate, hitherto insuperable barrier is not simply rebuffed, but rather, the question is transformed by rethinking its terms.

1 THE IDEA BECOMES A MACHINE THAT MAKES THE ART

Since the gradual withdrawal of the notion of practical skill as a necessary constituent of artistic practice it has been difficult to define with any degree of finality what it is exactly that an artist *does*, and what it is that constitutes the work of an artist, other than the tautological response: an artist is one who makes art.[1] Undoubtedly the most persistent answer, prevalent at least since the late 1960s heyday of conceptual art, is that the primary talent of the artist consists in the ability to generate ideas. However, this would seem to entail that the essence of the creative work must take place in advance of its construction and that the work of art always proceeds from a conscious plan rather than taking shape only in and through its enactment. In which case, the implementation of a "good idea" would be a mere formality, as no. 33 of Sol LeWitt's *Sentences on Conceptual Art* baldly states: "It is difficult to bungle a good idea."[2] On this model the physical manifestation is a mere vehicle and the artist must go about constructing it so as to maximize the clarity and impact of this animating principle. For instance, when considering the dimensions of a work, LeWitt writes, "the piece must be large enough to give the viewer whatever information he needs to understand the work and placed in such a way that will facilitate this understanding."[3] The enactment is a wholly pragmatic affair whose function is to expedite the transmission of its ideational content. However, is it possible to repudiate this radical pragmatism without overemphasizing or fetishizing the corporeal, sensual, or "performative" aspect of artistic creation? Carl Andre, often associated with conceptual art but vocally critical of its rhetoric, rejects the notion of the idea as sovereign. For Andre, "all ideas are the same except in execution. They lie in the head. In terms of the artist, the only difference between one idea and another is how it is executed."[4] And elsewhere he states, "if abstract art is art as its own content, then conceptual art is pure content without art. Following Reinhardt, I desire art-as-art, not art-as-idea."[5]

These questions surrounding the dialectical relationship between the ideational and the material (aesthetic) dimensions of art in the wake of conceptualism have recently been given fresh critical impetus by theorists engaging with the category of the "post-conceptual."[6] However, my central postulate in this opening chapter is that emergent neurotechnologies whose purpose is to automatically execute the intention or thought of the user cast this relationship in an entirely new light and call for a new theoretical framework. For if the idea is paramount, as first generation conceptual artists believed, would a technological bypassing of the corporeal negotiation involved in transmuting this idea into reality facilitate a more direct, true, or faithful realization? As John Dewey writes, "[the] act of expression that constitutes a work of art is a construction in time, not an instantaneous emission."[7] So if the artist is to have at her disposal a mechanism which precisely does enable the "instantaneous emission" of an idea it becomes necessary to ask how such a development would impact upon the act of artistic creation and to consider what *kind* of creation we would be dealing with. Neurotechnologies, in their idealized form, would not be one medium among others, which the practitioner must gain mastery over or at least obtain a degree of competency with. On the contrary, they promise the possibility of bypassing such skillful negotiation altogether, with the outcome being directly produced by a mental process. After all, even Duchamp's readymades involved a minimal degree of practical engagement with specific materials.[8]

Too Much, Too Little: Beckett and Joyce, Cage and Boulez

Kant writes in his *Anthropology* that "actuality is always more limited than the idea that serves as a pattern for realisation."[9] Undeniably, the process of physically realizing an artistic idea can often be experienced by practitioners as something of a struggle, fraught with obstacles and impasses. This imposing of formal limitations on what appeared boundless, and the conferral of clunky materiality onto the seemingly perfect product of the mind is the reason why the physical art object may disappoint its creator(s) and fail to live up to the promise of its gestation. It can seem that this leap from idea to object is an abyssal one, where the very first gesture feels like a compromise and a corruption. All further work undertaken as the artist attempts to bring it closer to the original conception is a form of regression to an irretrievable origin that is simultaneously a progressive dislocation from the origin, as the more she works the further it strays from its starting point. Like a palimpsest, it retains the traces of all that has been added and removed, such that even if the artist undoes what has been done and attempts to start afresh it is at risk of becoming labored, and referring more and more to itself and the previous failures than to the virginal idea in its original form.

This, of course, is in many respects how a work of art takes shape and becomes both more and less than what it may have started out as. Great things come from accidents or even from failing to accomplish what was initially intended. It can, however, have the opposite effect, and repeated frustration can stymie the creative process. This tension is given tragicomic expression by the eponymous narrator of Samuel Beckett's novel *Molloy*:

> This is one of the reasons why I avoid speaking as much as possible. For I always say either too much or too little, which is a terrible thing for a man with a passion for truth like mine. And I shall not abandon this subject, to which I shall probably never have occasion to return, with such a storm blowing up, without making this curious observation, that it often happened to me, before I gave up speaking for good, to think I had said too little when in fact I had said too much and in fact to have said too little when I thought I had said too much. I mean that on reflection, in the long run rather, my verbal profusion turned out to be penury, and inversely. So time sometimes turns the tables. In other words, or perhaps another thing, whatever I said it was never enough and always too much.[10]

The work of the artist is *never enough* in that more could always be done to a piece; perhaps she has not successfully articulated what she wished to express. It is *not quite there,* and the threat of misinterpretation or misconstrual is ever-present. There are innumerable accounts of artists being almost neurotically reluctant to declare a work finished and hand it over to be exhibited where it would circulate beyond their control, and as Alberto Giacometti once remarked, " [that's] the terrible thing: the more one works on a picture, the more impossible it becomes to finish it."[11] However, it is simultaneously *always too much,* in that the very first act—breaking the silence, sullying the clean white canvas or making the first mark on an empty page—is itself an excess, corrupting the purity of the idea or intention while committing the artist to that particular course. Every decision after the first breach takes her further away from where she originated, and following the path along the way she came never leads back to where she began; wiping away traces always makes more traces of its own.[12]

Another way the too much/too little axis can be approached is from the perspective of the viewer: what we are presented with is in itself *not enough*; it is too enigmatic and opaque. This gives rise to the need for some explanation or historical detail of what has been done and what it means, culminating in the increasingly widespread availability of explanatory audio guides at major art galleries, conveniently summarizing and deciphering these strange objects for our easy digestion. However, it is equally *too much* in that it places limits on our imagination and binds it to this particular object we are considering, limiting the possibilities to this particular work and nothing else. In other words, like the horror movie maxim that an unseen menace is always more terrifying than anything that could be directly shown, the more we are presented with the less freedom there is for the imagination. If one finally comes to experience a rare or hard to come by film or piece of music that one has long anticipated it can very often fail to live up to one's expectations. So the same thing that elevates our imagination to new heights simultaneously limits it, leaving us unsatisfied and impatient to move on to something else that will perhaps satiate us more. So if the artist inhabits the role of Beckett's Molloy, the position of the viewer as I have characterized it could, taken to its extreme, be identified with another character in twentieth century comic literature: the inscrutable master of disguise Henry Burlingame in John Barth's satirical novel *The Sot-Weed Factor*. He loves and loathes the world for this very reason—that while it inspires his imagination and "'twas splendid here and there . . . he could not but loathe it for having been *the case*."[13] That it is *the case* is both the cause and the limit of our enjoyment, grounding it and rendering it commonplace.

The artist caught in the double bind of "never enough and always too much" can often find herself tempted to, like Molloy, "give up speaking for good." She is thwarted by the seeming impossibility of creating a work of art and finds that the only available action is inaction, recalling Foucault's formulation of madness as "the *absence of an oeuvre*";[14] everything is compromised so how can one do anything? One strategy of coping with this impasse would be to do *almost* nothing, what the composer George Brecht called "borderline" art, involving barely perceptible events and interventions. Many of the evanescent practices of conceptual art, Fluxus, and the latter period of great

American and European abstract painting exist in tension with their own nonexistence. More recently, it is not uncommon in free improvisatory music for performances to hover just over the threshold of inaudibility: one strains to discern the body of a guitar squeaking as it is rubbed, or the slight rattling sound of a saxophonist's fingers on the keys. It is as if any breaking of the silence must be fully justified, and the longer it endures the more loaded and impactful any sound would become.

What this could not fail to call to mind is of course John Cage's famous composition 4'33". The piece is in three movements consisting of only one instruction: "tacet," the word used in western notation to tell a player to remain silent during a movement. As is well known, the title 4'33" is the total length in minutes and seconds of its notorious first performance in 1952 by the pianist and electronic composer David Tudor, who marked the beginning and ending of each movement by closing and opening his piano lid respectively; but Cage makes clear that the piece could last any length of time and be "performed" on any musical instrument or combination of instruments. Thus he imposes a wholly arbitrary and variable time frame during which what takes place is the "piece" but what occurs immediately before or after this duration is not.

4'33" is an "oeuvre," in Foucault's sense, reduced to the minimal difference between itself and its own absence: a limit inscribed in time that inevitably bleeds and undermines its boundaries. In stripping virtually everything away all that is left is the frame, and that is a wholly arbitrary and unstable one. Whenever a piece of music is performed the "start" and "end"—what is within and without of the work—is difficult to define but in Cage's piece this forms the very object of the work: some token signal from the performer acting as the only indication that the piece is under way, and even this is inferred rather than specified in the score. The clearly defined duration of the piece both excludes and includes the other side of that limit: time itself in its infinite expanse. In Derridean terms it *delimits* a certain space while *de-limiting* and contaminating that space in the very act of establishing it.[15] It excludes this boundless excess by drawing up the frame in the first place, delineating *this* period of time as the duration of a piece of music. However, as Kant notes, any "determinate magnitude of time is possible only through limitations [put] on a single underlying time. Hence, the original presentation, *time*, must be given as unlimited."[16] So in the very act of enclosing a section of time it evokes time itself, a sublime gesture that in Kant's terms negatively presents the absolute. The piece is almost nothing, threatening to collapse entirely but as such, as Derrida would have it, such a gesture merely draws attention to the instability of *all* limits and all acts of framing.

Beckett once famously said this of James Joyce:

> I realised that [he] had gone as far as one could in the direction of knowing more, [being] in control of one's material. He was always adding to it; you only have to look at his proofs to see that. I realised that my own way was in impoverishment, in lack of knowledge and in taking away, in subtracting rather than adding.[17]

In this is an implicit acknowledgement that one can *never* have complete control over one's production and attempting to do so will be an endless, Sisyphean task, for no matter how tightly sealed the frame may appear to be it will always allow for escape and infiltration. So Beckett may be said to approach this knot of too much/too little from the side of too little and Joyce that of too much. An argument could be made that these competing tendencies constituted the two poles of modernism, and that the foundering of the modernist project could be ascribed to each approaching its endgame—in the one case of there being nothing left to strip away and in the other complete saturation. Perhaps an even more exemplary illustration is the deeply ideological rift between the Experimental Music of Cage, Morton Feldman, Cornelius Cardew, and colleagues, and the classical avant-garde of Schoenberg and Webern through to Boulez, Stockhausen, and others. While the former were concerned with processes and chance procedures and treated notation as instructions for actions to be carried out, the results of which varied hugely from performance to performance, the latter attempted to rigidly define and contain what took place in their work. As Michael Nyman describes in his canonical study of Cage and his followers,

> [This] is the effect that processes have in experimental music: they are the most direct and straightforward means of simply setting sounds in motion; they are impersonal and external and so they do not have the effect of organising sounds and integrating them, of creating relationships of harmony as the controlling faculty of the human mind does.[18]

And at the opposing end of the spectrum,

> [One] finds Boulez, seemingly disconcerted by the impermanence of his sounds, constantly trying to fix them with ever greater precision by obsessive

> revising, refining and reworking, in the hope of sculpting his sounds into more permanent finality.[19]

For Nyman, the work of Boulez and his contemporaries in the European avant-garde was rigid and lifeless, demanding a less autocratic authorial approach. Boulez himself on the other hand would have considered the work of the Cagean school to be lacking in complexity and compositional rigor and needing more from the guiding hand of the composer.

Whereas traditionally musical notation was seen as an, albeit imperfect, method of transcribing a "musical thought," in the words of Edgar Varése, the Experimental Music composers, as noted above, all but abandoned conventional notation in favor of ambiguous worded instructions and pictographic diagrams that gave the performers unprecedented freedom of interpretation. This transformation of the score from a graphic *representation* of an arrangement of sounds into a set of *instructions* retroactively allows us to, as Nyman notes, see "the note C in a Mozart piano sonata [to mean] 'hit that piece of ivory there, with that force and for that long.' "[20] This echoes Willem de Kooning's famous declaration, "The past does not influence me; I influence it"[21] and similar sentiments expressed by Borges and Cage himself.[22] For the European avant-garde however, unconvinced by such radical moves, the conventional system of notation was still something to be struggled with, as an inadequate but nonetheless essential vehicle for transcribing one's "musical thoughts." This inadequacy became ever more pronounced when greater emphasis began to be placed on timbre than on pitch, and "nonmusical" elements began to be incorporated such as electronic devices and tape collages.

Stockhausen, who himself experimented with novel forms of notation, but for rather different reasons than those of the Anglo-American Experimental Music composers, once said,

> My intuition, my fantasy, is so much more developed than what I can do. I am constantly hitting my head against a wall because everything is so slow. I have so many ideas, and really brilliant ones, that I can never realise. . . . I will try to get a mission to a place which is more advanced, where music is far more developed. A place where I can work directly with the vibrations, where the atmosphere is adapted to the kind of vibrations that exist, where the people are so sensitive that I can make music right out of my consciousness, and where the translation is not so primitive. I have to go beyond writing, and using erasers that don't work, bad paper and all this. You see, it's all very primitive here.[23]

While Stockhausen has a well-deserved reputation for eccentricity and mysticism and comments like these should be taken in such a context, it nevertheless gives voice to a common frustration. This notion of creating work directly "out of one's consciousness" is an age-old dream of the creative artist. Raymond Scott, the composer and electronic music pioneer, expressed similar sentiments when he said in 1949,

> Perhaps within the next hundred years, science will perfect a process of thought transference from composer to listener. The composer will sit alone on the concert stage and merely *think* his idealised conception of his music. Instead of recordings of actual music sound, recordings will carry the brainwaves of the composer directly to the mind of the listener.[24]

This, in effect, is a yearning for the artistic idea to spontaneously appear fully formed, bypassing the crude practicalities and "primitive" media involved in the process of its enactment and eliminating all obstacles between conception and execution: the idea born without the aid of the midwife.

The earliest notable example of music produced directly from the brainwaves of the performer is Alvin Lucier's landmark 1965 piece "Music for Solo Performer." Developed in collaboration with the physicist Edmond Dewan, whose early research into brainwave control technology we discussed in the Introduction, the piece involved the performer (initially Lucier himself) doing nothing other than sitting on a chair with eyes closed wearing electrodes attached to his scalp. The amplified alpha waves recorded from the brain were channeled to loudspeakers with the resulting vibrations triggering a variety of percussion instruments placed next to the speakers. As Lucier described, "in this piece electronics allow you to go directly from the brain to the instruments, bypassing the body entirely."[25] However, while the sounds here are directly activated by brain activity this is evidently not music that has been thought into existence in the sense Stockhausen and Scott intend. Particularly so as alpha waves are only detectable in the brain when the subject is in a meditative, relaxed state of calm contemplation, and so the solo performer of Lucier's piece can only generate sounds when all intentional thought is reduced. It thus demands that he disengage from everything around him and not allow himself to be distracted by audience noise or passing thoughts. Again, in the words of Lucier, "[the] harder you try, the less likely you are to succeed; so the task of performing by not intending to, gave the work [a pleasing] irony."[26] The wish being expressed by the two composers above, on the other hand, is for a rather different sort of brain-generated music. Rather than the sonification of brainwaves, it expresses a yearning for their actual subjective states to be harnessed and for the music "in their head" to be automatically exteriorized as a fully-formed sound-object. As fantastically implausible as this may have sounded at the time the statements were made, it strikes us as considerably less so today in the wake of the extraordinary neurotechnological advances outlined previously.

A fascinating subfield of BCI research, known as brain-computer music interfacing (BCMI), is investigating different ways of translating intentional brain states into music for live, real-time composition and performance. The most notable figure in this area is UK-based Brazilian composer Eduardo Reck Miranda, founder of the Interdisciplinary Centre for Computer Music Research at Plymouth University. Miranda and his team have been working for over a decade on different methods to enable users to exercise meaningful control over musical interfaces using EEG signals. In a 2006 paper Miranda outlines the results of a system designed to interpret EEG patterns to give the user control over an acoustic piano.[27] Using a set of generative rules, identifiable patterns of EEG signals are assigned to a particular musical style based on a store of presets, which is then used to trigger the hammers on the piano to perform a piece of music in that style. So certain brain states will produce a piece in the style of Schubert, others Satie, and so on. The users can thus modulate their brain signals to produce particular styles but they have only very limited control over the outcome: it is a case of choosing with the mind rather than playing as such. Rather than deciphering and executing a particular note or succession of notes that the user intends, it is the computer that makes the compositional decisions, based on neural triggers. Miranda has subsequently conducted a series of brain scanning experiments geared toward understanding the neural correlates of music cognition, with a view to discovering patterns of brain activity suitable for BCMI control. In particular it is focused on analyzing active, attentive listening states versus passive, inattentive listening states so that by interpreting brain signals during audition the computer can determine whether a subject is actively attending to a piece, or part of a piece, as opposed to listening passively. Based on these outcomes Miranda has developed a prototype BCMI where the user is given the ability to actively shape and edit the music being generated in real time by focusing attention on specific constituents.[28] The piece can thus be transposed or reversed, notes can be added or removed, and the rhythm can be altered. However, while this is undoubtedly a greater degree of control than previously, the musical phrases are still automatically generated and the user is still not able to create the piece with the mind from scratch.

An alternative BCMI approach which aims to give the user greater control in creating and modifying a melody in real time from the ground up is to utilize what is known in neuroscience as the P300 response. This, as we described in the Introduction, is a fluctuation in the brain signal associated with an external stimulus that is detectable in EEG readings. The way the P300 response is typically exploited in BCI research is to present the user with a set of symbols on a computer monitor which correspond to specific actions that can be performed with the BCI. The users make their selection by focusing attention on the symbol corresponding to the relevant action, eliciting the P300 response in the brain which in turn prompts the computer to execute the desired action. Successful studies have been carried out by researchers at Goldsmiths College, London, and the Georgia Institute of Technology aimed at adapting this method of BCI control for musical composition and performance, where the symbols on the screen correspond to musical notes that are then identified and executed by the computer software.[29] So by attending to particular visual stimuli a rudimentary piece of music can be produced by intentional thought alone. This is evidently somewhat closer than Miranda's procedures to the ideal of realizing truly mind-generated music, although it carries limitations of its own, the main ones being that with the speed of present technology and the laborious nature of the task the piece produced is likely to have more scientific interest than musical.

Research into brain-computer interfacing for music is still in its infancy but the extraordinary results already achieved mean that we can now say with confidence that Stockhausen's dream of producing music "right out of [one's] consciousness" will almost certainly be realized in one form or another within the coming decades, whether through some combination of the above techniques or new methods yet to emerge. Gone, apparently, will be the necessity of the struggle with the instrument or the need to transcribe one's ideas into a tangible form ("writing, and using erasers that don't work, bad paper and all this"). It suggests that ultimately a lack of musical ability or instrumental skill need not necessarily be an obstacle to creating and performing music; one need only concentrate on the succession of notes, and the technology will be able to decode the neural activity and bring the intended piece forward into physical existence. Of course, the degree of physical activity varies greatly from one medium to another and is in no way correlated to the level of authorial involvement. In each case however, whether the tool employed is a piano, a paintbrush, a laptop, or a digital camera, certain technical and practical considerations must be negotiated and the resultant object will always contain both more and less than the authorial "idea" (however fully realized or provisional this latter may be.) What neurotechnologies seem to promise, if taken to their logical ideal, is an instrument without instrumentality: a technology that enables us to short-circuit the detour involved in the physical execution of an intended act.

Evidently, in the examples of BCMI techniques we have discussed, or of BCI technology more generally, it is not a case of imagining a particular mental content and then that content being automatically realized out of whole cloth. As Miranda writes, "the task of decoding the EEG of a person thinking of a melody, or something along those lines, is just impossible with today's technology."[30] It is more like techno-psychokinesis, whereby rather than physically executing an intended task it can be accomplished through thought alone. The mere intent to commit the action or bring about the event automatically and instantaneously coincides with its actualization. Thus the distinction between what one *can* do, and what one merely wants or intends to do but lacks the ability or the means to carry it out, becomes obscured. In the case of BCI-generated music, a user without any instrumental ability could conceivably compose and perform a piece of music simply by thought. While it has always been the case that a composer may write a piece for others to perform that she need not be able to perform herself, she will need a firm training in classical composition in order to internally hear how the marks on paper will sound when performed. The possibility raised by BCMI technology, however, is for *anyone*, regardless of conventional training, being granted the means to create a piece of music. This raises the question of whether the mode of expression (compositional know-how or instrumental skill) is something that could be learned subsequently in order to transcribe an idea that otherwise would find no outlet, or whether it is an essential constituent of those ideas. Can one truly have anything to say before knowing in principle *how* to say it?

A different form of exteriorizing mental content is being developed in the cognitive imaging experiments under way at Jack Gallant's laboratory at Berkeley that we encountered in the Introduction. Gallant's team has been able to show that live, dynamic perceptual experiences can be decoded from MRI patterns and digitally reconstructed on a computer screen. So unlike with

BCI procedures, in this case the very phenomenological contents of the mind are made manifest and displayed electronically. The methods employed were designed to reproduce specific visual stimuli from the brains of perceiving subjects, but it inevitably suggests the possibility of dynamic *quasi*-perceptual experiences such as imagining, dreaming, or hallucinating being similarly decoded and materialized. In fact, as Michio Kaku reports, describing a tour he made of Gallant's laboratory, such developments are already under way:

> Not only can this program decode what you are looking at, it can also decode imaginary images circulating in your head. Let's say you are asked to think of the Mona Lisa. We know from MRI scans that even though you're not viewing the painting with your two eyes, the visual cortex of your brain will light up. Dr. Gallant's program then scans your brain, and flips through its data files of pictures, trying to find the closest match. In one experiment that I saw, the computer selected a picture of the actress Selma Hayek as the closest approximation to Mona Lisa. Of course, the average person can easily recognize hundreds of faces, but the fact that the computer analyzed an image within a person's brain and then picked out this picture from millions of random pictures at its disposal is still pretty impressive.[31]

Impressive indeed, and what will be of concern for the remainder of this chapter are the possibilities for the creative utilization of neurotechnologies such as these and whether their use would radically transform artistic activity. For supposing that a procedure along these lines could be standardized and made commercially available suggests the possibility of simply sidestepping all of those difficulties that the artist often encounters when making the physical object equal to the thought behind it. If in the future we could coordinate cognitive imaging technologies, or a BCI controlled computer graphics system, with rapidly improving 3D printing techniques, which enable users to automatically manufacture solid objects from digital files, there is the very real possibility that an idea or imagined object could be automatically materialized. So we may now be approaching a situation where the technological tools available to the artist could enable her to bypass physical expression altogether.

Jack Gallant, speaking in a press release about his research, said,

> [At] the moment, when you see something and want to describe it to someone you have to use words or draw it and it doesn't work very well. You could use this

> technology to transmit the image to someone. It might be useful for artists or to allow you to recover an eyewitness's memory of a crime.[32]

Evidently he regards his system as a more efficient and effective means of exteriorizing one's interior space; that drawing or "using words" are clumsy and inadequate for such purposes and are hence destined to be superseded by more advanced techniques. However, in the case of a work of art, matters are considerably more complicated. An artwork is not merely an outward manifestation of a prior mental concept, where the means employed to execute it can be substituted for other means with no bearing on the outcome produced. In the words of Dewey, the work of art is "a construction in time, not an instantaneous emission." So at issue is what takes place *in between*—the gap between thought and actuality, and between what the artist intends to do and what he or she actually does—and whether this is at least as integral to the work as the motivating idea itself.

Knowing and Doing: Kant and the Creative Act

Up until now we have been guilty of oversimplification in forcing a rather crude separation between (internal) idea and (external) expression, the latter tasked with bringing the former to fruition. This has served a heuristic purpose, but it is ultimately unsatisfactory as an account of artistic creation. Turning now to Kant's account of the creative act in the *Critique of Judgement* will help us to flesh out this picture and develop a more nuanced depiction of what takes place in the creative act, which in turn will allow us to see more clearly the implications of neurotechnological advances for creative practice.

First, works of art "in general" are distinguished from nature "as doing (*facere*) is from acting or operating in general (*agere*); and the product or result of art is distinguished from that of nature, the first being a *work* (*opus*), the second an effect (*effectus*)."[33] The act of *doing*, (or *making* as the James Creed Meredith translation renders it), implies a decisiveness and deliberateness that is not to be discovered in natural processes. The *work* of art, contrary to the *effect* of nature, must be the result of an intentional consciousness that has decided upon a course of action in advance. Prior to the object's existence there was the will to act; it did not occur as a result of simple causality, serendipity, or blind instinct.

Second, "*Art*, as human skill, is also distinguished from *science*" in the same way that we would distinguish *I can* from *I know*, or the practical from the theoretical. "That is exactly why," Kant says, "we refrain from calling anything art that we *can* do the moment we *know* what is to be done" (303). So although there must be an anticipatory presentation of the intended result on the part of the artist, which is what separates it from a natural process, this idea or intention is not sufficient in itself. "Only if something [is such that] even the most thorough acquaintance with it does not immediately provide us with the skill to make it, then to that extent it belongs to art. Camper describes with great precision what the best shoe would have to be like, yet he was certainly unable to make one" (303–4).[34]

It is therefore not merely the capacity to conceive of the object but the practical ability to bring about its existence that distinguishes the artist from the nonartist. Put differently, it is the *know-how* rather than the *know-that*. A critic or philosopher may claim to know what constitutes a great work of art, but only the artist can put this knowledge into practice.[35] However, with neurotechnologies such as brain–computer interfaces or cognitive imaging techniques, the distinction between *knowing* and *doing* collapses. To know or think it *is* to do it. Contrary to the suggestion that neurotechnology amounts to nothing more than an increased efficiency in the means by which we exteriorize our thoughts, Kant shows us that the implications would extend much further than simple convenience. Far from being a mere shortcut to the same outcome, the very essence of creative activity would be transformed. Given Kant's description of artists as being those with the ability to enact or perform that which others may only be able to conceive, neurotechnologies would accordingly promise, to a greater or lesser extent, to endow anyone with the capacity to create a work of art. The crucial aspect, however, may turn out to be the status of this *greater or lesser extent*.

The latter point we have dwelled upon has been taken in isolation, its original purpose being only to differentiate art, as something that necessarily carries with it a practical component, from science, which allegedly does not. Furthermore, this is with regard to "art in general," of which *fine art* is a species, evidenced by the example cited being Camper's lack of shoemaking know-how rather than a lesser artist who lacks creative genius. The very idea of knowing fully and comprehensively what is to be done before carrying it out is anathema to artistic creation, to the experimentation and improvisation that takes place in the creative act. The Kantian genius works blindly to a large extent, in the absence of determinate knowledge of what he is doing.

If he *were* to know in advance and thus be able to provide a full account of what it is that he is doing (and why) it would imply that he worked to a rule or template. This has a bearing on our earlier discussion concerning the struggle involved in the realization of an idea, which Kant here describes as the "slow and even painful process of improvement, directed to making the form adequate to his thought without prejudice to the freedom in the play of those powers" (312–13). Above we depicted this struggle as the difficulty in bringing the physical form up to the level of the idea, thus placing the art object in a position of subordination. Here, however, it is a delicate balance between two forces: the imagination (on the side of the idea) and the understanding (on the side of form). What we have been rather loosely calling the artist's "idea" is equivalent to what Kant refers to at certain times as imagination and at others as genius. As Henry Allison has noted, Kant's account of genius is notably inconsistent: at times it is identified as virtually synonymous with imagination (what Allison calls the "thin" conception of genius) and at others as including the understanding and judgement along with the productive imagination (the "thick" conception).[36] In either case, whether genius includes understanding, or whether understanding operates "afterwards" to regulate the workings of genius, there is a negotiation required between the dynamic imaginative component and the formal presentation.

So not only must the object do justice to the idea, the idea must allow itself to be regulated by the understanding without sacrificing the play of its powers. The idea is not the unimpeachable sovereign, which the artist must endeavor to satisfy and avoid doing violence to; it also must compromise, otherwise the artist is at risk of producing "nothing but nonsense." Indeed, if there is a conflict between the two faculties and something must be sacrificed, "then it should rather be on the side of genius; and judgement, which in matters of fine art bases its decision on its own proper principles, will more readily endure an infringement of the freedom and wealth of the imagination than that the understanding should be compromised" (320). Working in harmony with the understanding, the imagination is brought down to earth but it can display the fruits of its wanderings by means of what Kant names aesthetic ideas. These are a counterpart to rational ideas, but whereas the latter are concepts which cannot be presented by any intuition, the aesthetic idea is a form which no concept can express. Rather than a failure of presentation to be equal to the thought, it is a failure of thought to comprehend the presentation. The artist "ventures to give sensible expression" to that which "goes beyond the limits of experience" (314), and therefore beyond what can be cognized and thoroughly explicated. Thus the expression is of equal significance to the imaginative conceit:

> Hence genius actually consists in the happy relation—one that no science can teach and that cannot be learned by any diligence—allowing us, first, to discover ideas for a given concept, and, second, to hit upon a way of *expressing* these ideas that enables us to communicate to others, as accompanying a concept, the mental attunement that those ideas produce. (317)[37]

So the talent of the artist consists jointly in discovering the idea *and* in presenting it so that the idea can be experienced by others. An aesthetic idea is not a mental event that must then somehow be presented or transcribed in a physical form but *is* the form itself. Through producing aesthetic ideas the artist incites the same feeling in the viewer of the work as he himself experienced in its discovery, thus bearing witness to a community of sense. This incitement, however, through which the artist enables the viewer to experience what he himself has experienced in its creation is what we might call indirect: mediated by the art object in which it finds expression. What the new neurotechnologies promise is for this shared experience between artist and viewer to take place *directly*, bypassing the intermediary stage just as Raymond Scott envisaged. We will pick up this theme again in chapter 4.

Intuition and Expression: Croce's Aesthetics

It is precisely the emphasis on the subjective harmony of cognitive faculties which forms Hegel's principal quarrel with Kant's aesthetics. For while Kant identified the beautiful as the point of harmonious union between universal and particular (the idea presented within the sensible on equal terms) this correspondence was located squarely on the side of the judging subject. Hegel remains close to Kantian aesthetic ideas but takes the critical step further in characterizing this union between idea and form as true and actual in itself and not as merely subjective. So while in Kant's account the idea remains infinitely out of reach, untouched by its presentation, and being merely evoked for the subject, in Hegel's aesthetic philosophy the idea must come forth of itself to be presented and "carries within itself the principle of its mode of appearance."[38] A purely general idea, abstract and undetermined, could never be presented, evoked, or even merely suggested, so "determinacy is, as it were, the bridge to appearance."[39] There is thus, for Hegel, no gulf to be bridged between the idea and its manifestation since the idea involves in itself its own means of sensible presentation. Similarly, for Gilles Deleuze it makes no sense to speak of a vague or general idea. Rather, ideas must be "treated like potentials already engaged in one mode of expression or another and inseparable from the mode of expression, such that I cannot say that I have an idea in general."[40]

This is where the philosophy of Benedetto Croce becomes important for us, which combines an idealist aesthetics with a phenomenalist account of perception that offers a particularly useful model for thinking through the relationship between internal and external in the wake of neurotechnologies. At its heart, Croce's aesthetics is a philosophy of intuition, where intuition is portrayed as a creative activity through which we impose form and order upon the raw sense data of experience. This intuition is a type of knowledge, independent of and prior to intellectual knowledge, which is based upon our power to form mental representations. Concepts are second-order abstractions based on these mental representations, so intuitive knowledge is necessarily primary. Thus if logic is the science of intellectual knowledge, aesthetics is the science of intuitive knowledge.

For Croce there is no reality prior to its constitution by our intuitive activity. All mental representations we form of the world function as a creative ordering, and in doing so we articulate this information and make it comprehensible to ourselves. From this active model of intuition Croce argues that "everything that is truly intuition or representation is also expression."[41] In order to make the sense data we receive cognizable we must give it form, and this is already expressive. Since expression is at work from the ground up the notion of an inexpressible, intangible thought, idea, or impression is inadmissible. We have not "had" an idea or thought until it has been expressed, before this it would be only a confused, muddled feeling. It is then merely a small and insignificant step to give a bodily form to this thought. This recalls Hegel's rebuke to the Romantic conviction that what is most inspired in the heart of the artist is inexpressible, and hence there exists an infinite depth to him beyond what his work discloses: "On the contrary, his works are the best part and the truth of the artist; what he is [in his works], that he *is*; but what remains buried in his heart, that *is* he not."[42] Since the idea is expression from its inception, its success or failure consists entirely in *how* it has been expressed.

Nevertheless, when the clumsy physicality of the work seems not to live up to the artistic cognition, even if the artistic rendering stayed perfectly faithful to it, the artist may feel disappointed as if all that imagined potential has been captured and grounded in this object. However, there may in fact be a perfectly mundane explanation for this: it simply wasn't as successful an idea as she believed it to be. It is only by testing one's ideas through the process of realization that we discover their true nature and, in the words of Karl Popper, "let our false theories die in our stead."[43] Since to have an idea *is* to express it, if we are unable to express ourselves clearly then this is because we have not fully grasped or developed our own thought. It follows that we could not say of a work of art that its expression does not do justice to its idea, as if it were simply a case of shabby clothing.

An immediate objection here could be to appeal to the countless examples in the history of popular music where a song has been recorded in several forms, many of which improve on the original rendition. Surely this bears witness to a kernel of unrealized potential that had not been fully expressed in the song's first performance. However, this would be to conflate two distinct meanings or even stages of "expression": the song itself, and its performance or recording. The Crocean response would be that the artistic *intuition-expression* proper is the song itself, and the execution is a technical matter only:

> So clearly are the two forms of activity distinguishable from each other that one could be a great artist but a poor technician . . . but what is impossible is to be a great poet who writes poor verses, a great painter who does not know how to match colours . . . in short, a great artist who does not know how to express himself.[44]

Proust, for example, was a notoriously poor editor and proofreader of his manuscripts, which led not only to grammatical mistakes but to awkward errors of continuity and verisimilitude, but these technical failings in no way detract from the beauty of *À La Recherche du Temps Perdu*. The notion of a great poet who lacks lyrical ability however, or a great painter with no eye for composition, is an oxymoron for they are the very matter of the expression itself.

However, in the case of popular music this raises further questions, for it appears to suggest that there is no creative artistry in, to take an obvious example, John Coltrane's performance of "My Favorite Things." The implication would be that Coltrane's contribution is mere technique with all the legitimate artistic work belonging to Rodgers and Hammerstein, which is plainly false. Croce, while not addressing this example specifically, would no doubt consider it to be a case of translation. Translation is an activity that Croce judges to be strictly speaking impossible, because one cannot extract the content of one expression and present it "in the guise of another."[45] Either the translation is slavishly faithful and hence a diminished version of the original, or:

> It creates an entirely new expression by putting the original expression back in the crucible and mixing it with the personal impressions of the one who calls himself the translator. In the first case the expression stays the same as it was originally, the other version being more or less inadequate, that is to say, not properly expression: in the other case there will indeed be two expressions, but with two different contents. (76)

Only an innovatively reimagined interpretation of a song would be a separate expression and hence a work of art in its own right rather than a derivative retread. However, this does not contradict the earlier claim that the song itself is the expression proper and the performance a mere technical matter because a creative cover version is not merely a more successfully produced repetition of the original but is a reinvention of the song itself.

Croce decisively rejects the notion, postulated by Kant, of the layperson who may know what constitutes a successful work of art but lacks the skill to execute it. The difference between a layperson and a great artist like Raphael is not simply the latter's skill in rendering the world in paint. Given Croce's insistence that there is no reality prior to its construction by our intuitive activity, it follows that the artist quite simply inhabits a different world to the nonartist. "The painter is a painter because he sees what others only feel or glimpse but do not actually see" (10). His power of representation is greater than ours and hence his power of expression is greater than ours, since they are one and the same. While we may imagine our sense perception to directly acquaint us with things as they are, what we in truth experience are indices of a book: "the labels that we have attached to things and which take the place of them" (10). These indices are perfectly sufficient for our everyday lives but are quite inadequate for the artist's purposes. After all, the usual foundation and point of departure for life-drawing classes is to teach the students *how to see*. The amateur drawer's ineptness comes not (or not only) from a lack of skill with the pencil or charcoal but because one has to learn to perceive the world differently in order to make a compelling two-dimensional rendering of it.

So to the question as to whether neurotechnology would make artists out of all of us, the proper Crocean response would be to say: not any more than we *already are* all artists. The important point to be made is that the difference between the aesthetic activity of the artist and that of the rest of us is not one of kind but only of degree. No technology could miraculously unleash latent creative potential, liberated by the removal of the necessity for physical or technical skill of any kind. The *intuition-expression* of the artist, poet, composer, and so forth, will always be richer than that of the nonartist, thus it is nonsensical to imagine a mechanism that would find the "hidden talents" of those who were unaware they had them, because creative ability is not something buried within us that awaits activation: it is *already expressed.*

All of this leaves us with the question of what exactly is the status of the art object itself in Croce's model if all the work of the artist takes place internally prior to the production of any physical traces. As Croce writes,

> The aesthetic stage is completely over and done with when impressions have been worked up into expressions. When we have captured the internal word, formed an apt and lively idea of a figure or a statue, found a musical motif, expression has begun and ended: there is no need for anything else. That we then open, or want to open, our mouths in order to speak, or our throats in order to sing, and, that is, to say aloud and full throat what we have already said and sung sotto voce within ourselves; and that we stretch out, or wish to stretch out, our hands to touch the keys of a piano, or take a brush or chisel, following, so to speak, on a larger scale those small and swift actions which we have already executed, translating these into a material where traces of them will remain more or less durably;—all this is something additional, and obeys quite different laws from those governing that earlier activity, laws which we are not for now concerned with: although from now on we recognise that the latter activity produces objects and is practical or voluntary. It is customary to distinguish between the work of art which exists inside us and that which exists in the outside world: this way of speaking seems infelicitous to us, since the work of art (the aesthetic work) is always internal; and what is called the external work is no longer the work of art. (56–7)

Within Croce's description is a curious intersection with pure conceptualism. It strongly recalls such conceptual mantras as this from Sol LeWitt's "Paragraphs on Conceptual Art": "when an artist uses a conceptual form of art, it means that all of the planning and decisions are made beforehand and the execution is a perfunctory affair."[46] The physical embodiment given to an intuition is nothing more than an aide-mémoire; it is a material supplement to something that has already taken place, something that was alive in the mind of the artist. As Walter Benjamin famously wrote, "[the] work is the death mask of its conception."[47] If the rendition is successful then the physical entity will incite or reproduce in oneself and in others an already experienced or produced intuition. Much like what Derrida has written about the impossible ideality of the archive, in Croce's account the successful work of art "*comes to efface itself*, it becomes transparent and unessential so as to let the *origin* present itself in person. Live, without mediation and without delay."[48]

The work of art always refers to something that has passed and is nothing more than a reproduction. It is technically incorrect, according to Croce's account, to call a work of art beautiful; rather, it is "simply an aid to the reproduction of internal beauty" (114). This chimes with Mozart's description of his creative procedure, in which the idea arrives and then "the committing to paper is done quickly enough, for everything is already finished."[49] There is a further correspondence with biographical accounts of Brian Wilson's compositional process. It is well documented how he mentally conceived his "pocket symphonies" down to the smallest detail, and the exacting standards he demanded of the session musicians, involving endlessly repeated takes, were required in order to bring them to recreate what he could hear internally.[50] Of course, what can never be ascertained is how far away from the original conception he may have traveled during this process. It may be that he believes himself to be remaining faithful to the mental image while imperceptibly modifying it with each performance. As LeWitt states, "the fewer decisions made in the course of completing the work the better. This eliminates the arbitrary, the capricious, and the subjective as much as possible."[51] The less an artist leaves himself to do in the execution, the purer the idea can be and the less interference there will be between execution and the final result. This interference is, of course, reduced to the absolute minimal degree by neurotechnologies and so on this account, they seem to promise an unprecedented level of purity and faithfulness to the idea.

It is clear that Croce's model, or that of LeWitt's conceptualism, is the complete antithesis of performative accounts of artistic creation. It is difficult to square Croce's theory of art with free improvisation or aleatoric practices in which there is supposedly nothing prior to the act: no notion of an outcome, everything begins and ends with the doing. There is something automatic and spontaneous about such practices that cannot be the result of a prior, deliberate mental construction. In free improvisatory musical performance a great deal of the skill resides in the ability to react to one's fellow players and follow the piece where it leads quasi-organically. However, if we try and remain within Croce's logic we may want to problematize this notion of expression and insist on there always being a prior intuition, no matter how minimal. All great improvisers, from Alice Coltrane and Ornette Coleman to Derek Bailey and Evan Parker, practice long and intensely precisely so that "in the moment" they can instantly bend their instrument to their will and elicit any desired sound or motif. This, it could be argued, is simply in microcosm what the musician-composers have always done: they have something in mind that they intend to play and then aim to make the instrument produce it. Improvisation is sometimes referred to as "instant composition" and in light of this we could say that it is still the same model of composition but the stakes are higher because the music is being played out live with no opportunity for revision. As such, mistakes, happy accidents, and inadvertent inspiration abound, which lends to the improvised music a sense of being more "alive" than a similar piece that has been composed and rehearsed. So the interval between the (Crocean) intuition and the act may shrink to an infinitesimal gap, but the former does not dissolve entirely into the latter. The suggestion that an improvisatory pianist, such as John Tilbury, would blindly hammer away at the keys with no notion of what he is doing is obviously false and does a disservice to his art.

In fact Croce anticipates this potential objection to his depiction of the physical art object as a reproduction:

> To that theory we have outlined of physical beauty as simply an aid to the reproduction of internal beauty, that is, to expression, it could be objected: that the artist creates his expressions in the act of painting and sketching, writing and composing; and that, therefore, physical beauty, rather than coming after, can sometimes come before aesthetic beauty. That would be a superficial way of understanding the procedure of artists, who, in fact, do not even make strokes of the brush without first having seen by means of the imagination; and, if they have not yet seen, make brushstrokes, not to externalise their expressions (which do not then exist), but as if to try out and to have a simple point of support for their internal meditations and contemplations. (114)

It would be absurd to suggest that the painter Joan Mitchell, for example, "intuited" in advance her large, expressionistic compositions and worked the paint on the canvas in order to recreate what has already taken place internally. But it would be equally wrong and, again, do her a gross injustice if we were to assume that it was an entirely random procedure with no aesthetic guidance. This aesthetic guidance determines whether certain gestures made are accepted or rejected, according to a particular vision of what she wants to achieve. If the standard by which she judges were to be taken from elsewhere, according to some preconceived model, the results would be derivative. But if this standard comes from *within* then we have not yet entirely left Croce behind.

Such arguments notwithstanding, it is very easy to see that such a comprehensive and totalizing philosophy of art as Croce's, in which artistic labor is treated as a homogeneous activity, is no longer a tenable proposition. Moreover, it cannot be denied that the process of art making entails a great deal of investigation, both material and conceptual, and often the work travels of its own accord far from whatever the artist initially set out to achieve. The idea develops with and through the execution; after all, to look at Picasso's preparatory studies for *Les Demoiselles d'Avignon* alongside the finished work is not to see successive attempts at realizing a stable and unchanging idea but is to see the maturation of the idea through its materialization. So we can no longer, as Croce seeks to do, insulate "internal beauty" from "physical beauty." As John Dewey writes,

> [Between] conception and bringing to birth there lies a long period of gestation. During this period the inner material of emotion and idea is as much transformed through acting and being acted upon by objective material as the latter undergoes modification when it becomes a medium of expression.[52]

In cognitive psychology this necessity of the externalization process in the formation of the "internal idea" has been empirically investigated. The authors of a 1999 study set out to examine why a painter needs to have recourse to multiple sketches rather than simply picturing the final work internally and then executing it on canvas.[53] Their claim, based on the findings of their study, is that human apperception is constrained in ways that it is not in direct perception. It is exceedingly difficult, so they say, for us to construct an internal image that is interpretatively open, while this is obviously something that a work of art demands. We tend in our mental imagery toward fixed understandings that do not allow multiple interpretations, so the artist relies on perception to supplement the shortcomings of her imagination. As Andy Clark notes in commenting on this research, "the iterated process of externalising and re-perceiving turns out to be integral to the process of artistic cognition itself."[54] This artistic cognition, as Katherine Hayles has argued, is an embodied cognition, and includes within it the interactions with the materials and the many conscious and unconscious bodily movements that participate in its creation.[55]

Croce even admits this possibility in the passage we have just cited, where he concedes that artists sometimes start to work *before* completely conceiving a piece, in which case they will "make brushstrokes . . . to have a simple point of support for their internal meditations and contemplations." However, this ever-present possibility for the external act of painting to influence and support the purely "internal meditations" already corrupts the purity that Croce is concerned to maintain, introducing an economy of forces to the ostensibly autonomous intuition.

The artistic cognition, or Croce's intuition, is thus reliant on its supplement, the material support, in its very formation. Furthermore, if every aesthetic intuition is already expressed then it follows that it must be expressed *as* something, of which the intuition is the internal anticipatory visualization. This is something Croce accepts but he is not able to see that if the aesthetic intuition derives its content from the medium, which after all is a contingent historical product, then this fatally undermines the ontological primacy of the idea over the physical object. For if the idea is a mental image *of* the intended outcome then the material object itself can no longer be said to be something secondary, additional, or inessential. It is an obvious point that the possibilities open to any artist at any time are conditioned by the technological affordances available to them. Suffice it only to recall Heinrich Wölfflin's famous dictum: "Not everything is possible at all times, and certain thoughts can only be thought at certain stages of the development."[56]

So what we can retain from this discussion of Croce is the conviction that an idea is always already its own expression, but we cannot follow him in his attempts to restrict the whole of this process to the interior of the artist's cognition, disavowing the effect of the material support on the internal process. So while in Croce's model the essential incommensurability between internal and external may have been overcome, this is only achieved by giving one term absolute priority over the other. Everything is placed on the side of the idea, with the result being that the material object or event itself is thoroughly diminished. However, if the idea is born and develops only through interactions with the technological medium we would have to conclude that the artist's mind is only one component in the formation of the idea. This would be to read Croce against himself and in doing so bring him into alignment with Andy Clark and David Chalmers's notion of the extended mind, where our cognition (or intuition) is distributed between the brain and its technological and social supports.[57]

Given the inevitable transformation of the idea during the process of materialization, it seems clear that this is where any enquiry must be focused that sets out to consider the implications of the artistic utilization of neurotechnologies. In effect, this crucial limit between idea and execution described by Dewey is apparently removed from the picture. Anecdotal accounts suggest that very often it is an accidental, unplanned, or unforeseen element that turns out to be an artist's most cherished part of a work, something that could not have been anticipated until it stands before them (which is supported by the abovementioned psychological study), and it is difficult to imagine how such serendipitous by-products could result from this mode of creation. It would seemingly be nothing more than an *intention* plucked straight from the mind with no physical negotiation. So in negating the corporeality and the improvisation, not to mention the struggle, involved between conception and birth (to continue Dewey's metaphor), would the result be a stillbirth?

Improvisation and Intuition: Beyond the Instrument

Slavoj Žižek likens the advancement of new media to the Hegelian negation of negation, which Žižek defines as a shifting of coordinates. His example is how digital computers were first introduced as a means of more efficiently going about our activities, such as writing and printing newspapers, but before long they came to render obsolete the very activities they were designed to make more efficient.[58] So if neurotechnologies were to shape how we produce and consume art, the art object itself may soon become an unnecessary intermediary, which could be bypassed altogether in the form of direct brain-to-brain communication. The aesthetic intuition in Croce's sense would thus be communicated directly from the mind of the artist to the mind of the receiver, as Raymond Scott envisaged, without the need for the physical "memory-aid."

This echoes Marshall McLuhan's contention that the "new medium is never an addition to an old one, nor does it leave the old one in place. It never ceases to oppress the older media until it finds new positions and shapes for them."[59] McLuhan's claim is borne out by the all but total replacement of analog technology with digital in film, photography, and music and the reduction of analog forms to lifeless fetish-objects. However, rather than immediately giving rise to radically expanded forms of expression the potential of the new media is invariably constrained by our tendency to "force [it] to do the work of the old."[60] Pushed to its extreme this tendency leads to such kitsch absurdities as artificial tape hiss, fake vintage photographs, or laser projected virtual keyboards that reproduce the sounds of a physical keyboard for authenticity. With this in mind we should consider neurotechnologies in terms of what Clement Greenberg would call their medium specificity, meaning their affordances and their limitations. At first glance, what seems difficult to envisage is how spontaneity could be maintained in work produced via brain–computer interface, since it apparently demands an *explicit, deliberate* act of thought to induce the apparatus to perform the intended action.

Stockhausen has recounted how during his early musical training music was considered to be solely a rational activity, with its intuitive aspect entirely suppressed. While his early compositions could accurately be termed "rational music," realized through intellectual energy, in his later compositions he endeavored to create a space for a form of free improvisation, liberated from the conventions and idioms of jazz or any other tradition, that he named "intuitive music." In this spirit, his piece *IT* consists of a textual score which simply instructs the performer to "think NOTHING. Wait until it is absolutely still within you. When you have attained this begin to play. As soon as you start to think, stop and try to re-attain the state of NON-THINKING. Then continue playing."[61] Undoubtedly, it is a requirement of all forms of freely expressive improvisation for performers to, in a sense, disengage their conscious attention and perform without reflecting on what they are doing. This is true of any automatic motor activity, from table tennis to touch typing; if one starts to think about what one is doing it is as if a spell has been broken and what before came naturally and with ease now becomes awkward and labored. This is the difference between what cognitive scientists call the implicit and the explicit systems of learning. A receptionist may be an expert typist but be unable to recite from memory the order in which the letters appear on the keyboard. Similarly, many proficient musicians find that they are incapable of teaching others to play the instrument because this would require what has been learned implicitly to be relearned explicitly, which demands a considerable expense of effort. Implicit learning is learning-through-doing and takes place largely in the absence of explicit knowledge about what has been acquired, and any amount of transfer of skill from implicit to explicit control during the act will be highly detrimental to its performance. Stockhausen says of intuitive music that, "acting, or listening, or doing something without thinking, is the state of pure intuitive activity, not requiring to use the brain as a control."[62] However, with music created by neurotechnology the brain is the *only* control, so it seems that of the two paradigms Stockhausen describes only the former, rational music, would be possible with such a system.

However, interestingly, a recent work by London Fieldworks, a collaborative, interdisciplinary art project established by Bruce Gilchrist and Jo Joelson that worked with BCI technology makes the very act of not thinking, or thinking of nothing, the generative force behind the piece of work. Titled *Null Object*, the project involved Gustav Metzger, the German artist known primarily as the originator of Auto-Destructive Art in the 1960s, attempting to think about nothing for long stretches of time while an EEG machine monitored his brainwaves. The information recorded from Metzger's brain while thinking of nothing was translated into a three-dimensional shape, which was subsequently excavated from the inside of a cubic block of limestone (dating from the Jurassic period) by a robotic drill, creating a hollow void. The three-dimensional shape information was based on a sort of brain database, consisting of EEG data from hundreds of participants, whose brains were scanned while perceiving depth information within random-dot autostereograms (made famous by the *Magic Eye* books in the 1990s.) The correlation between the preset shapes perceived and the brainwaves of the perceiving subjects was the basis for the translation of Metzger's brain signals into shape information, which was then hollowed out from the stone. The exhibited work, shown at the Work Gallery in West London in late 2012 to early 2013, consisted of the stone itself along with supplementary material including a film documenting the drill in action, a framed image and scale model of the resultant shape, and schematic diagrams of the whole process.

This is a piece of work that is subtractive rather than positive: the absence of thought made manifest by negative space. Metzger's involvement, almost as medium rather than as participant, resonates with his own interest in negativity and subtraction, culminating in his famous three-year art strike of the late 1970s. So rather than an object being thought into being as we have been envisaging, this is the visualization, through absence, of the very absence of thought. However, it is still the making manifest of the purely psychical:

something brought about through thought alone, but not through the thought of the event triggering its actualization because here there is no "object" of thought to be realized. Obviously issues of authorship are paramount, and rather than being the product of a single author or limited collective it is the outcome of an open-ended network of actors, including Gilchrist and Joelson, Metzger, those hundreds who donated their brainwaves to the database, the psychophysicists who pioneered the use of random-dot autostereograms (primarily Béla Julesz and Christopher Tyler), the engineers and scientists who developed the technology that made it possible and all of the preceding research that their work built upon, the technology itself, and so on. The thought or idea of the artist in this work is exponentially reduced—not only by virtue of the fact that the artist thinks of "nothing," but that the artist himself whose thoughts are harnessed is not the sovereign originator of the work but merely a component in the system.

This piece, as well as others by London Fieldworks utilizing neurophysiological interfaces offers an enthralling indication of the potential creative uses of neurotechnologies.[63] However, it must still surely be considered as belonging to the paradigm of a "rational" rather than "intuitive" practice as Stockhausen describes. For, while Metzger may be thinking of nothing, this is nevertheless a concerted act and the resulting object bears only an indirect, figurative relation to the thought that triggered it. In this respect it strongly recalls Lucier's "Music for Solo Performer," which also used the act of thinking of nothing to set in motion an elaborate technical system. In neither case is the performer "improvising," as his task is strictly delineated in advance, and the outcome produced, while subject to a certain degree of chance, is not the result of some immediate, intuitive inspiration.

The renowned guitar player Derek Bailey identified two distinct improvisatory attitudes towards the instrument, which he dubbed pro- and anti-instrument. For the former, the instrument is the be-all and end-all of the performance, and the player is merely there to tease out all of its capabilities and sonic potential. This leads to modifications, preparations, and extensions of the instrument, all by way of doing full justice to its possibilities and respecting its limitations. The practice of London Fieldworks could loosely be described as a pro-instrument utilization of neurotechnology due to the reduction of the role of the motivating thought of the artist and the central function played by the technology itself. Their work exists at the vanguard of what is technologically possible in an art performance and could not even have been conceived before the technology it utilizes was available.

On the other hand, as Bailey suggests, "the anti-instrument attitude might be presented as: 'The instrument comes between the player and his music.' . . . Technically, the instrument has to be defeated. The aim is to do on the instrument what you could do if you could play without an instrument."[64] Here the instrument stands in the way between the performer and her expression and is fundamentally an inconvenience, or even an irrelevance, since it is the player and her inspiration alone that counts. So it should make little difference *what* the instrument is, since it is merely a conduit, which is why the instruments favored by such performers usually have an extremely limited capacity—the idea being that a more direct expressiveness is enabled when the opportunities for technical virtuosity are minimized. On the one hand, BCI technology seems to offer such performers the fulfillment of this ideal, representing the ultimate victory of expressiveness over the tyranny of the instrument. As suggested above, this is potentially an instrument unlike any other, for with it the very instrumentality, physicality, or materiality of the mediating vehicle is seemingly effaced. On the other hand, however, the demands of its operation may prohibit such immediate inspiration from finding an outlet.

However, the suggestion that neurotechnologies such as BCIs could only respond to explicit, deliberate acts of thought reflects a familiar misconception about the nervous system, expressed in the commonly used but misleading term "muscle memory." All somatic motor commands, whether explicit or implicit, issue from the motor cortex, so to suggest as Stockhausen does that intuitive practices bypass "the brain as control" is incorrect. The pertinent question is whether through practice users of a brain-controlled system could be afforded a comparable degree of fluency as professional pianists and trombonists have with their instruments, and there would appear to be no intrinsic reason why the users could not. After all, any new implicitly learned skill is initially performed deliberately before becoming effortless and automatic. Even the kind of NON-THINKING meditative state that Stockhausen demands of his performers is a skill that must be purposively learned: witness the recent boom in mindfulness training as evidence. The composer Gordon Mumma says, recounting his experience of performing Lucier's *Music for Solo Performer*, that "the process of non-visualising . . . is a specially developed skill which the soloist learns with practice; and, no matter how experienced the soloist has become, various conditions of performance intrude upon that skill."[65]

As Andy Clark notes, describing performance artist Stelarc's famous prosthetic third hand, "[what] we are witnessing, in Stelarc's fluent performances, is yet more evidence of the remarkable capacity of the human brain to learn new modes of controlling action and to rapidly reach a point where such control is so easy and fluent that all we experience is a fluid, apparently unmediated mesh between will and motion."[66] While his auxiliary limb was only "indirectly controlled by neural signals from Stelarc's brain," being operated by muscle activity in his legs and abdomen, BCI technology enables *direct* neural control of the machinery promising even greater fluency and expressivity. To return to the experiment discussed in the Introduction where laboratory monkeys learned to feed themselves using a brain-controlled robotic arm, the mechanical arm soon became assimilated into unconscious activity and integrated into its normal behavioral routine, and this occurred much more quickly than any of the scientists anticipated. Effectively this is the difference between explicitly thinking "now I am going to move the arm" and doing it automatically as if it were an extension of the operator's body. Already, in the embryonic stages of development of BCI technology for musical performance, reports suggest that the interface itself becomes less and less obtrusive with use, in an analogous way to when one reaches a certain level of skill with an instrument and can play without having to think about what one is doing.[67] This promises an even truer fulfillment of Stockhausen's ideal of intuitive music: unimpeded expression emanating automatically, bypassing the body as well as all conscious control.

With this we find a response to our earlier question about whether the transformation of the idea in and through the process of exteriorization, what Clark called "the iterated process of externalising and re-perceiving," would have to be sacrificed in a practice making use of neurotechnologies, as all practical negotiation would effectively be bypassed. On the contrary, it is easy to see how the artistic cognition could still develop and remodel itself on the basis of its automatic realization in exactly the same way as the artist working through sketches or models, but here the process would be radicalized. For if the physical outcome is automatically and instantaneously synchronized to the thought process behind it the artists can, in effect, encounter their aesthetic intuition as physically instantiated and see it take shape and transform live as their intuition evolves—*as a result of that externalization*—in a fluid, reciprocal exchange. Simultaneously "internal" and "external," the intervening gap between concept and execution would indeed be overcome but not at the expense of the artist's ability to revise and develop the initial guiding idea. So whereas in Croce's model the interval between the internal intuition and its technical manifestation is overcome only by completely subsuming the art object into the idea, here the gap is short-circuited without subordinating either term of the relationship to the other. As in Croce's philosophy, the intuition, or artistic cognition, includes its own expression, but without this whole process being confined to the interior of the artist's mind.

However, just what kind of creation is this, where thought and physical event coincide as one, quasi-immediately? This is a question we broached at the outset of the chapter and it remains a crucial one. Moreover, it is evidently a much broader question than one concerning artistic production alone. We are asking what it would mean for human agents to be able to directly manipulate external reality via thought alone, or for the interior contents of the mind to be immediately transposed outside as an object. Slavoj Žižek addresses this question in relation to BCI technology in his 2006 book *The Parallax View*. In particular, he considers the laboratory monkeys that learned to feed themselves using a BCI-controlled mechanical arm and a related study with human participants who played a computer game with a BCI which had been programmed to recognize motor activity in the brain corresponding to hand movements of the joystick. From this Žižek draws the bold conclusion that what neurotechnologies accomplish is nothing less than the overturning of human finitude. In the next chapter we will take up this provocative claim which, characteristically, Žižek does not develop any further but instead leaves as a tantalizing, almost throwaway remark.

2 INTELLECTUAL INTUITION AND FINITE CREATIVITY

The claim that technology will make—or has already made—Gods of men is an old and extremely familiar one. The supposition is that humankind has so greatly upgraded its biological capabilities with technological enhancements that they now resemble powers which previous generations would have ascribed only to (the Judeo-Christian) God.[1] Perhaps the most noteworthy articulation of this claim is made by Freud in *Civilization and its Discontents*, where he famously writes that due to our advanced technological and cultural acquisitions "man has, as it were, become a kind of prosthetic God."[2] The tacit assumption underlying Freud's observation is that technological progress occurs in a continuous, uninterrupted sequence of augmentations—"with every tool man is perfecting his own organs"—with the ultimate goal being the overcoming of finite limitations in the divine attributes of omnipotence, omniscience, and omnipresence, not to mention immortality.[3] Moreover, it conceives of the relationship between God and man as existing on a single, graded plane with God as the upper extremity.

So when Žižek suggests that neurotechnologies will bestow God-like capabilities upon the user the temptation might be to view it as simply one in a long line of prosthetically enhanced God-man narratives. However, this would be to fail to grasp the true stakes of Žižek's extraordinary proposition, which is noteworthy for two important reasons. The first is that the God invoked is Kant's Supreme Being, and not the anthropomorphic, or superhuman, God of popular imagination. For Kant the human and the divine structurally obscure one another and share no common territory. The second reason is that the ability to directly intervene into the external world with the mind, or to automatically bring into existence an object of thought, does not simply amplify the functioning of an existing organ or faculty but instead constitutes an entirely original power. The question is what sort of power this is and what the consequences of exercising it would be. Žižek writes:

> Even Steven Hawking's proverbial little finger—the minimal link between his mind and outside reality, the only part of his paralyzed body that Hawking can move—will thus no longer be necessary: with my mind I can directly cause objects to move; that is to say, it is the brain itself which will serve as the remote-control machine. In the terms of German Idealism, this means that what Kant called "intellectual intuition"—the closing of the gap between mind and reality, a mind-process which, in a causal way, directly influences reality, this capacity that Kant attributed only to the infinite mind of God—is now potentially available to all of us, that is to say, we are potentially deprived of one of the basic features of our finitude. And as we learned from Kant as well as from Freud, this gap of finitude is at the same time the resource of our creativity (the distance between "mere thought" and causal intervention into reality enables us to test the hypotheses in our mind and, as Karl Popper put it, let them die instead of ourselves), the direct short circuit between mind and reality implies the prospect of a radical closure.[4]

There are three significant claims here, which need to be considered separately. The first is that a form of cognition characterized by Kant as belonging only to the mind of God is made available to us by neurotechnologies. The second is that this capacity would enable us to overcome our finitude. The third is that this very overcoming of finitude, rather than bestowing upon us an enhanced power of unconditional freedom would in fact see us relinquish the very form of creativity that we possess. However, the reverberations of such an event would not only be felt by future acts but would have retroactive impact as well, suggesting that human finitude, and creativity, must be reconsidered from the ground up. So this "deprivation" would be total and irreversible, and the genie cannot be returned to the bottle once it has been opened.

Assessing the potency of this claim demands that we pass through a careful reading of the notion of intellectual intuition in its precise Kantian formulation, foregoing the transformations and modifications the term subsequently underwent in the hands of Fichte and Schelling.[5] Finally, we will consider how and in what respects our freedom as well as our creativity springs from the finite gap between internal and external, such that if we surmounted or overcame this constitutive finitude through technological advances our freedom would be put radically in question.

However, the task of providing a coherent, univocal definition of intellectual intuition (or the intuitive or archetypal intellect which would possess such a faculty) is a particularly tortuous one. All the references Kant makes to it—and it appears in all three *Critiques* as well as numerous other works—are fleeting and episodic, dealing with problems that at first sight appear heterogeneous. Thus it is a slippery, multifaceted concept.[6] This is no doubt due to the fact that it is exclusively used negatively, or problematically. Indeed the main thematic treatments of it in the *Critique of Pure Reason* were only contributed in the second edition, which provides a clue to its function as a buttress to the system, but as a limit rather than a positive addition. It operates primarily in order to demonstrate that were we to transgress the limits of this system we would immediately involve ourselves in contradictions and absurdities. Thus it is invoked merely to ground our finite cognition and root it in the senses, and it is introduced only to be denied as a possibility.

The theme of the intuitive intellect as a faculty of the infinite mind of God appears in Kant's work as early as the *Dissertation on the Form and Principles of the Sensible and the Intelligible World* of 1770, where he writes, "the divine intuition [is] the cause, not the consequence, of objects."[7] Two years later, in the letter to Marcus Herz in which he famously introduces the scope of the critical project, he returns to it with regard to the correspondence between the object of our representation and the representation itself: What, he asks, guarantees the reference of the internal representation to the external object? If the former is merely the result of the subject's being affected by the latter then it can be explained as of cause to effect. Otherwise, if the representation was "*active* in its relation to the object—"i.e., *if the object were produced by the representation itself* (as one thinks of divine cognitions as the archetypes of things)"—then we could likewise understand the congruence between internal and external, "so one can at least understand the possibility of both an *archetypal* intellect, upon whose intuition the things themselves are grounded, as well as an *ectypal* intellect, which attains the data of its logical activity from the sensuous intuition of things."[8]

From these brief readings it already starts to become clear what Žižek has in mind and why he could be led to assert that such a divine faculty is "potentially available to all of us" in the wake of neurotechnology. The representations in the mind of God are characterized as being the ground or cause of the objects themselves, while the opposite is the case for human cognition. However, what neurotechnologies facilitate is for a purely psychical event to have real effects in the outside world and even for it to be made manifest as an object, thereby transforming our passive representations into the active ground of physical events. The constitutive passivity of our cognition, its reliance on the givenness of external data, thus seems to be technologically overcome.

Gegenstand/Ent-Stand

Kant's system of transcendental idealism is developed in opposition to "dogmatic" metaphysical realisms founded on the epistemic premise that we can know entities as they are in themselves based on what we know of them from experience. It tacitly assumes that things as they appear to us are the same as they are outside of any reference to our faculties of cognition. God's vantage point is here posited as the limit case, with our human perspective being but a lesser, although strictly homogeneous, derivative. On this model the further we extend our understanding of nature the closer we come to infallible knowledge. However, Kant's gesture is to radically demarcate finite cognition from infinite cognition, in the process placing insuperable limits on human knowledge.[9]

The first lines of the Transcendental Aesthetic, and hence of the *Critique of Pure Reason* proper, state that intuition is the only means by which our cognition refers directly to an object, while thought acts indirectly, gathering and ordering the data that has been intuitively given. This occurs, however, only through the givenness of the object, which, "for us human beings, at any rate" is brought about exclusively through hetero-affection.[10] There are thus two features to our faculty of intuition, the first and primary factor being that the objects preexist and condition the act of intuition, hence it is *receptive* and not *productive*. Secondly, since thought cannot directly access this data it must take place through the sense organs, therefore it is *sensible* rather than *intellectual*, and we can have no knowledge and no means of accessing objects without their being first intuited in this way:

> Our kind of intuition is called sensible because it is *not original*. I.e., it is not such that through this intuition itself the existence of its object is given (the latter being a kind of intuition that, as far as we can see, can belong only to the original being). Rather, our kind of intuition is dependent on the existence of the

> object, and hence is possible only by the object's affecting the subject's capacity to present. . . . For the reason just set forth, intellectual intuition seems to belong solely to the original being, and never to a being that is dependent as regards both its existence and its intuition (an intuition that determines that being's existence by reference to given objects). (B72)

As seen in the passage from the letter to Marcus Herz, Kant identifies only two possible bases for the relationship between our intuition of an object and the object thus intuited: one must be the ground of the other. If intuition were itself understood to be the ground this would succumb to the material idealism of Berkeley, whereby objects only exist in and for an act of perception, so it must be the case that the object is the cause of the intuition. Our intuition is capable only of receiving something given, and since all thought only aims at the material so given, all knowledge is entirely dependent on these given objects, which exist in their own right. However, Kant also says something more than this: we are dependent on the receptivity of our intuition not merely for our knowledge of the external world but also in terms of our very *existence*, that existence being determined "by reference to given objects." This is intended as a further defense against material idealism, whether of the "dogmatic" kind associated with Berkeley or the "problematic idealism" of Descartes, the latter asserting that absolute certainty belongs only to inner experience, meaning the existence of external reality can only be inferred and never proven. What Kant means to demonstrate is that it is only by reference to the objects of outer intuition that we are able to determine our own existence, thus any doubt or outright denial of the existence of an external world is precluded.

The claim is that in order to connect and retain the successive moments in the flux of consciousness we need to be able to refer to an enduring substrate. This cannot be a mere representation within me if it is to be something against which I measure such subjective representations, so inner experience is possible only against the backdrop of external, actually existing objects. The intellectual consciousness we have of our existence, the bare *I think* of transcendental apperception, is nothing without sensuous intuition. It thus does not follow from this *I think* that *I am*. If, on the other hand, "with that intellectual consciousness of my existence I could at the same time link a determination of my existence through *intellectual intuition*, then this determination would not include necessarily the consciousness of a relation to something outside me" (Bxl). Only a being in possession of such a faculty could determine itself through the mere thought of itself, because that act of the intellect would, as a result of its immediate access to its "object (the self), simultaneously create it and all its determinations (*causa sui*) and would not be dependent upon external objects.[11] Precisely because we are not in possession of such a faculty we can only gain knowledge even of our own selves through the receptivity of intuition and the independent existence of external objects. So Kant's rejoinder to the empirical idealist or the skeptic is to say that without objective outer experience, which to them is doubtful, there would be no *inner* experience, which one cannot doubt without performative contradiction. However, it appears that Kant only secures this defense against the wrong kind of idealism at the cost of upsetting the distinction between the transcendental and the empirical. I am not simply the constitutive, transcendental subject if my constituting act depends upon constituted objects in order to function. There is a circularity here that Bernard Stiegler will exploit in his reading of Kant in the third volume of his *Technics and Time* series. He argues that this necessary recourse to externality bears witness to our originary prostheticity and our reliance upon what he calls "tertiary retention," or material memory traces.[12]

So receptivity is primordial. As Heidegger points out, our intuition is not receptive simply because we happen to be endowed with the particular sense organs that we have, rather it is only because we are in the midst of beings that are not of our creation and not under our power that we must be in possession of organs of sensibility through which these beings can be given to us. Our discursive cognition is divided between a receptive and a spontaneous faculty, this spontaneity functioning only to synthesize given sensible data. Without the unity bestowed by thought intuition would be merely a "blind play of representations" (A112) and the *I* would dissolve amidst the confusion of sense-impressions. So "*receptivity* can make cognition possible only when combined with *spontaneity*" (A97) and this is how thought brings the play of representations to objective knowledge of objects.

It is precisely this productive, spontaneous nature of thought that explains why an intuition that is intellectual could not be receptive. Rather, it must be a form of cognition that would render its object immediately present through the mere act of thinking it. This is because the understanding, for Kant, does not derive its concepts from experience but rather brings them to experience. If thought were not reliant for its objects on their being given from outside but could instead engage with them immediately, the only way this relation could be conceived is if thought was itself the ground of those objects. Our intuition is thus doubly finite: dependent both upon the object being given rather than freely creating that which it intuits, and upon that intuited object being determined by thought since it is not transparent to the act of intuition. The infinite mind of God could not conceivably be dependent upon objects to which it has to conform because the Supreme Being could have no such limitations. Its objects must therefore spring forth from the cognition itself:

> Absolute knowing discloses the being [in the act of] letting-stand-forth and possesses it in every case "only" as that which stands forth in the letting-stand-forth, i.e., it is disclosed as a thing which stands forth [als Ent-stand.] Insofar as the being is disclosed for absolute intuition, it "is" precisely in its coming-into-being. It is the being in itself, i.e., not as object. Strictly speaking, therefore, we do not really hit upon the essence of infinite knowledge if we say: this intuiting is first produced in the intuiting of the "object."[13]

So we think naïvely if, in an anthropocentric way, we conceive of God's relation to "objects" as an emission, springing "out" from the mind in the way that we would imagine such an ability in ourselves. That is, it could not be conceived as something *inner* that takes on substantial being *outside* the mind since where the infinite mind is concerned there could be no "outside" against which it could define an "inside," for externality itself (space) is a form of finite intuition. Once this being of thought would have taken on objectivity then it would no longer be under the mind's power and would exist in its own right, regardless of how it came into existence.[14] As Alva Noë has remarked, the idea of an infinite power of sensible intuition (vision in this case) is one that can have no sense for us since to experience a sensible entity is nothing other than to "come up against its limits and one's own."[15] To see an object is to be guided by what that object offers to us: if it is opaque or transparent, flat or many sided, and so on. To see it from all possible angles at once, and to see its contours while simultaneously seeing through it is not to encounter it at all. So the infinite intuition of God cannot relate to objects existing outside of that act of intuition as to do so would be to experience its own limitations. The arising in and for the act of intuition is the mode of being that this *Ent-stand* would have to have. Our knowledge, as finite, "necessarily conceals" the thing in itself, for in being given as object [Gegenstand] it is impossible for "that same being to be seen as that which stands forth" [Ent-stand.][16]

However, here we are faced with a problem which Kant seemingly cannot address, namely, of accounting for our world of appearances from the perspective of the Supreme Being. If things in themselves remain entirely under God's power and their mode of being is identical with their arising in thought then what does God know of spatiotemporal determinations? The only way of resolving this is by making the division, so to speak, symmetrically asymmetric, so that not only is it the case that we, as finite beings, cannot know anything of the things as they are in themselves, also from the vantage point of the things in themselves the emergence of appearance cannot be accounted for. Each "side" must structurally obscure the other. However, this is as much as to say that there is something, namely the finite world of appearance, which is outside of God's power, which can only undermine His status as supreme. If to resolve this situation we say simply that it is a consequence of our restricted view of the things in themselves—as they *really* are—then we return to a precritical metaphysics that treats the finite as merely the limited perspective on the infinite; but finitude cannot be incorporated into the infinite as a relation of part to whole. In short, if we *start* from the side of the infinite there is no way of accounting for how the finite comes to be, so for Kant we can only start from the side of the finite. The things in themselves can be conceived only as regards how they must be determined *for us*; the infinite cannot be thought in itself but only in its *relation* to the finite. What this leaves unasked and incapable even of being posed is the question of finitude itself: why finitude? The only way this question could be addressed is by leaving finitude behind, in a Hegelian move, and accounting for it as the self-limitation of the absolute, but for Kant this would be wholly illegitimate. More than once Kant insists that we cannot ask *why these categories and no others,* or *why do we have the particular form of intuition that we do,* because here we encounter the unsurpassable limits of our knowledge. However, in an ironic reversal, this very limitation of our knowledge ends up privileging our perspective over that of the Supreme Being. From "our" side, presupposing finitude as already given, we can account for things in themselves, as the necessary substrate of appearance, and to a certain extent we can determine them as such. Conversely, from the *other* side, finitude is entirely inexplicable.

Possibility and Actuality

Kant demonstrates that, to avoid a crude skepticism which denies the possibility of discovering universal necessity, or the unphilosophical empiricism that would derive the latter from experience, we must assume that the regularity and binding necessity of natural laws is brought to experience by our powers of understanding rather than being discovered therein. Since the understanding is spontaneous and its concepts are not derived from experience it is easy to fall into the illusion that it is independent of intuition, while the latter is wholly dependent upon the former. However, this would be to forget that the understanding is nothing other than a faculty of rules for ordering the sensible manifold and without that data it is completely empty. This is why Heidegger characterizes the understanding as even more finite than intuition, because intuition is primary and provides the sole immediate access that we have to beings, while understanding is indirect and must refer to something general in order to return to the particular. If we were to imagine an intuition undetermined by thought something would at least be given, whereas a faculty of combining the manifold can signify nothing at all if no manifold has been given. So due to the irreducibly finite nature of thought, "all [God's] cognition must be intuition rather than *thought* which always manifests limits" (B71).

If we were to suppose that the understanding could in fact determine its object *immediately* and perform its function without having to rely on sensible intuition to provide it with its raw materials, this "object" cannot have been given to it from outside as this could only come about through the senses. Rather, it would be freely given by and for itself. However, such an understanding would bear no relation to our discursive faculty, for "if I were to think of an understanding that itself intuited (as, e.g., a divine understanding that did not present given objects but through whose presentation the objects would at the same time be given or produced), then in regard to such cognition the categories would have no signification whatever" (B145). Nevertheless, because these rules of the understanding are not gathered from the senses we are naturally led into the fallacious belief that they are independent from sense and that purely logical reasoning can provide insight into an intelligible world of reason (*mundus intelligibilis*). However, once again this could only be the case for an understanding that intuited, for without intuition it has no capacity to access objects. Therefore, the grounding condition of sensible intuition, that it is reliant upon already existent objects to be given, is *a fortiori* true for the understanding.

Nevertheless, the idea of the noumenon for Kant cannot simply be cast aside because we find ourselves unable to know anything of it. Since the objects of experience are only appearances, there must be something underlying this appearance that is not accessible to experience. To be finite means that our mode of cognitive access to the world is neither absolute nor the only possible kind. And because it is only for beings like ourselves that the conditions of sensibility hold then this underlying substrate cannot be extended in space nor will it have a temporal duration. Thus the understanding can conceive of the supersensible but only negatively, by abstracting from the conditions of sensible intuition. If, on the other hand, we were to think a positive concept of noumena, as the "*object of a nonsensible intuition*" (B307), we would need to assume an intellect that is capable of accessing this other dimension. This must be a faculty of intuition, because only intuition provides direct access to objects, but since the "object" here is nonsensible, this intuition must therefore be an intellectual one. The noumenon then, like intellectual intuition, is a problematic concept, serving only to indicate the boundaries of our finite cognition. In this way our understanding can venture beyond its proper sphere, but only problematically: the actual *existence* of this world can in no way be determined by the mere concept. However, indicating the boundaries beyond which our knowledge cannot reach still amounts to an extension of knowledge, albeit a negative extension.

The concept of a purely intelligible entity is an entirely undetermined one because we cannot think of any way in which such an entity could be given to us. Indeed, as Heidegger drew our attention to above, the notion of a noumenal *object* is contradictory. As Kant writes,

> We cannot call the noumenon such an *object*; for this signifies precisely the problematic concept of an object for a quite different intuition and for an understanding quite different from ours—an object that hence is itself a problem. Hence the concept of the noumenon is not the concept of an object; rather, it is the problem, linked inevitably with the limitation of our sensibility, as to whether there may not be objects wholly detached from this sensibility's intuition. (A287-8/B343-4)

The unknown thing called noumenon is not an *object* since objecthood is the result of an act of synthesis performed by our cognitive faculties. The sense data that we receive through intuition is not given as unified, so the impression of unity can only be an act of our intellect. This synthesis does not apply only to the chronology of our experiential flux but accounts for the very wholeness or consistency of objects. Thus, corresponding to the transcendental unity of apperception is the concept of the transcendental object. This, likewise, is a purely formal point of unity, or frame, to which we refer the matter of sensation and whose role is to keep our cognitions from "being determined haphazardly or arbitrarily, [and ensure] that they are determined *a priori* in a certain way" (A104). This transcendental object is consequently unknowable, but not in the same way that the noumenon is unknowable, that is, entirely outside of all possible experience and having no relation to our faculty of cognition. It is unknowable because, having no content, there is nothing in it to know. It is nothing but the concept of an object in general, which as such sits at the limit between the inside and the outside of experience. If it fell on the inside then it would be just another piece of sense-data which itself would then need to be determined according to some other rule of synthesis. If it were outside of experience, like the noumenon, then it could have no determining influence on our cognition.

The distinction between *mundus sensibilis* and *mundus intelligibilis* assumes our senses and our understanding to refer to ontologically distinct regions when they can perform their functions only in concert. An intuition lacking conceptual determination, or a concept with no corresponding intuition would be merely an empty presentation without an object. If we were to conceive of an object of which we could have no intuition, and yet we wanted to assert its existence then this could only spring from the understanding itself. This, for us at least, is impossible because the understanding gives us access only to the possibility of a thing. For the actuality corresponding to this possibility some sensation must be given.

Kant defines the possible as being that which "agrees with the formal conditions of experience" and the actual as that which "coheres with the material conditions of experience (with sensation)." These "postulates" of any empirical thought do not in any way determine the object of experience itself and are not added to the concept of a thing as a predicate. As Kant demonstrates in his famous refutation of the ontological proof of God, existence is not a real predicate and cannot be inferred from the mere concept of a thing. No matter how far we explicate a concept we learn nothing whatsoever about whether it corresponds to a real entity without its being given to intuition. Rather than contributing to the concept or determining the object itself, the postulates express "only the object's relation to the cognitive power" (A219/B266), namely, whether the object is given to us as possible, meaning only to the understanding, or as given also to sense and therefore actual. On the side of the object itself actuality and possibility are entirely indifferent, they merely signify whether a possible perception accompanies the thought of the thing in the subject. Possibility signifies something "inside" us while actuality must refer to something "outside" us. This "outside" is contingent and beyond our powers to determine.

Such a finite limitation clearly cannot apply to the Supreme Being, which is not given but gives its "objects." This point is made clear in the *Critique of Judgement*, where Kant writes that it is only for a discursive cognition such as ours that the distinction between possibility and actuality holds sway. An intuitive understanding would be one in which the actual and the possible would coincide:

> Human understanding cannot avoid the necessity of drawing a distinction between the possibility and the actuality of things. The reason for this lies in our own selves and the nature of our cognitive faculties. For were it not that two entirely heterogeneous factors, the understanding for concepts and sensuous intuition for the corresponding objects, are required for the exercise of these faculties, there would be no such distinction between the possible and the actual. This means that if our understanding were intuitive it would have no objects but such as are actual. Concepts, which are merely directed to the possibility of an object, and sensuous intuitions, which give us something and yet do not thereby let us cognise it as an object, would both cease to exist. Now the whole distinction which we draw between the merely possible and the actual rests upon the fact that possibility signifies the position of the representation of a thing relative to our concept, and, in general, to our capacity of thinking, whereas actuality signifies the positing of the thing in its intrinsic existence apart from this concept. Accordingly the distinction of possible from actual things is one that is merely valid subjectively for human understanding.[17]

The important point to be made is that the ability to immediately instantiate the contents of the mind in material form that is bestowed upon us by neurotechnological enhancements would enable us to traverse at will this finite distinction separating "possible from actual things." It could not be altogether annulled because it corresponds to the unalterable structure of our cognition but it would surely be brought under our power to a truly unprecedented extent. The *actual* existent thing could be instantaneously brought about from, and out of, the merely *possible* thought of that thing.

Nevertheless, it is easy to see that this does not constitute an overcoming of finitude, for two reasons. First, because once brought into independent existence the actual object is no longer under the absolute power of the mind that generated it. This is why we said earlier that divine cognition could not be conceived as the production of an inner content that takes on material existence outside the mind. Second, because the possible is only apparently more under our power than the actual. A priori limits structure any possible object even before it is actualized, and we are not able to freely modify them through the power of speculation or arrive at a position from which to account for them (hence the ban on questioning "why the categories?" or "why space and time?"). So when the distinction between possibility and actuality is as modest as it is in Kant, the ability to traverse it at will is not a supreme power.

Any actual existent must agree with the form of possible experience in general. Thus form precedes matter, which reverses one of the fundamental doctrines of precritical metaphysics, which holds that the components must precede their arrangement. This, Kant shows, is because there is an assumption that human understanding has a direct insight into being, unconstrained by the conditions of sensuous intuition. Therefore, since the understanding requires for something first to have been given in order to then determine it, in this account unbounded, absolute reality is first given as the ground of all possibility, which is subsequently determined by limitations—space and time: "And thus it would in fact have to be, if pure understanding could be referred directly to objects" (A267/B323). Since this is not the case and we have knowledge only of appearances—which must conform to the conditions of sensuous intuition (space and time)—then the form must precede the matter, which is unthinkable for the "intellectualist philosopher." Therefore, on Kant's account, limitation does not come afterwards to determine a given infinite substance, rather, it is the irreducible condition of the possibility of experience of actual entities. Everything starts with finitude but since finitude cannot be absolute the realm of the noumenal thing in itself is posited as a necessary correlate of the thing as appearance. In the beginning there is not the thing in itself, generated through the infinite intellectual intuition of the Supreme Being, which is subsequently limited in space and time when perceived by human intuition. This would collapse the two heterogeneous aspects into one and treat our sensuous perspective as if it were merely a confused intellectual intuition. If finitude comes first, however, and God is only posited out of the necessity of stabilizing our phenomenal world of appearance, then once again we as finite beings seem to be placed *above* the infinite being.

The Imagination and the Schematism

As Heidegger shows, in the course of developing the distinction between possibility and actuality, "it turned out that the positing of the actual proceeds out of the bare concept of the possible, out into the outside, over against the inside of the subjective condition of the subject."[18] Because the finite conditions of possibility stem from *our* cognitive faculties, we project the form in advance to which the actual must conform in order to confront us as outside of us. This horizon, upon which alone beings can be encountered, is the guiding theme of Heidegger's early work. In his famous *Kantbuch* of 1929 this takes the form of a radical phenomenological interpretation of the transcendental imagination.

Kant defines the imagination in the second edition of the *Critique of Pure Reason* as "the power of presenting an object in intuition even *without the object's being present*" (B151). It is the first act of synthesis performed by the mind, which binds the sensible manifold into an image. This acts as a mediating faculty, falling between sensibility and understanding and enabling the latter to perform its conceptual operations. Before the understanding can act upon the given and bring it to unity, the prior synthesis of imagination must have taken place. There is, however, some confusion over the peculiar status of the imagination and whether or not it is to be positioned alongside sense and understanding as one of the fundamental faculties. The first line of the Transcendental Logic could not be more explicit:

> Our cognition arises from two basic sources of the mind. The first is our ability to receive presentations (and is our receptivity for impression); the second is our ability to cognise an object though these presentations (and is the spontaneity of concepts.) (A50/B74)

In the first edition version of the Transition to the Transcendental Deduction of the Categories, however, Kant writes,

> [There] are three original sources (capacities or powers of the soul) that contain the conditions for the possibility of all experience, and that cannot themselves be derived from any other power of the mind: viz., sense, imagination, and apperception. On them are based (1) the *a priori synopsis* of the manifold through sense; (2) the *synthesis* of this manifold through imagination; and finally, (3) the *unity* of this synthesis through original apperception. (A94)

This apparent inconsistency is underscored by what Heidegger calls the imagination's "homelessness," owing to the bipartite division of the doctrine of elements into the Transcendental Aesthetic and the Transcendental Logic corresponding to the two primary faculties. The reason for this disparity is that the imagination cannot simply and straightforwardly be said to be a separate source of knowledge whose elements and contribution to cognition can be examined in isolation, as with sensibility and understanding, but it performs an indispensable function that neither can do in its place. It is both receptive and spontaneous, performing a figurative rather than conceptual act of synthesis.

There are two distinct functions performed by the imagination, a reproductive and a productive function. The reproductive imagination, conflating under one name the dual operations that Husserl would later call primary and secondary retention, is responsible for the consistency of the time series. Without the ability to hold onto what has been given each new moment would wipe away the last and there would be no possibility of experience. Evidently there can be no a priori act of reproduction since it is entirely subject to empirical impressions. The productive imagination differs from the reproductive in that it freely and originally gives form to the content received from intuition. Like the categories of the understanding, the form thus bestowed is not taken from experience, precisely because there is no form to be so given until it has performed its function; hence it is spontaneous. However, because it deals directly with the matter of intuition, rather than indirectly via concepts, it is simultaneously receptive. This aspect-giving is not a subsequent act of assembly that is performed only once the disordered raw materials have been given, rather the material is formed in its very appearing. So if the reproductive imagination's power of retention is the enabling source of the past, the productive imagination projects ahead toward future impressions, and it is only through their joint holding-projecting (or retention and protention in the language of Husserl) that we can experience the consistency of the present.

There is both an empirical and a transcendental use of the productive imagination. Its empirical use is outlined in the *Anthropology,* where Kant describes it as "a faculty of the original representation of the object (*exhibitio originaria*), which consequently precedes experience."[19] However, as Heidegger stresses, this original representation is not to be conflated with the original creation of the Supreme Being:

> This original presenting, however, is not as "creative" as *intuitus originarius*, which creates the being itself in the intuiting. The productive power of imagination forms only the look of an object which is possible and which, under certain conditions, is perhaps also producible, i.e., one which can be brought to presence. The imagining itself, however, never accomplishes this production.[20]

In the wake of neurotechnologies, however, this distinction is apparently overcome, which is precisely why they seem to constitute such an unprecedented event. We stand to be in possession of a mechanism which in fact does allow the *imagining itself* to accomplish the realization of its objects directly. A BCI or cognitive imaging process equips the user with the ability to imagine or intend a physical event, action, or object and for it to be automatically brought about. So if the only difference between the *original exhibition* of the productive imagination and the *original intuition* of the divine intellect is the former's confinement to playing with forms of possible objects and its inability, unlike the latter, to bring about the presence of the actual object in reality then neurotechnologies do indeed open up the possibility of a fundamentally new faculty of mind corresponding to intellectual intuition.

However, in the very next paragraph of the *Anthropology*, just after Kant has outlined the powers of the productive imagination and its independence from experience, he immediately limits its scope, insisting that "[the] productive faculty, however, is nonetheless not creative, because it does not have the power to produce a sense impression which has never before occurred to our senses. One can always identify the material which gave rise to that impression."[21] The productivity of the imagination only amounts to a reorganization of sensibly given material and not true creation in the sense which we would ascribe to an infinite intellect. To take a concrete example, the music created through a BCI system will always be operating within a preexisting musical language that is being triggered by the mind rather than physically performed. Even if it were to so radically disrupt that language that it sounded to its hearers as if it were an entirely new vocabulary—like twelve-tone serialism for instance—thought does not by itself and out of nothing give "the existence of its object" as the faculty of intellectual intuition must. This is why even the most inventive work of science fiction remains recognizable and familiar and cannot escape earthly concerns, introducing creatures, intelligences, and societies that are identifiably of our world. Kant himself gives the example of artistic renderings of Gods and angels, which always take human form because we are unable to imagine any other form for a rational being to take. Of course this is not a despondent diagnosis and does not amount to the denial or downgrading of originality. Rather, it is only to affirm that originality is strictly finite and can only operate within limits that it did not freely determine itself. This is nothing more than to say that poets do not with every new poem invent an entirely new vocabulary from the ground up but must work within the constraints of their given language. The production of the new from out of the familiar is exactly what originality is, and as with Kant's account of aesthetic ideas, the artist of genius can use this sensibly given material to produce something that causes the mind to transcend the sensible, but she cannot present such supersensible ideas directly.

This inability to create out of nothing is not only a subjective constraint but is also an ontological condition, for the very consistency of our experience rests on there being no rupture in the time-series. As such, "creation cannot be admitted as an event among appearances, because its very possibility would already annul the unity of experience" (A206/B251). If something arose out from nothing then there would have to have been a point of time in which it was not, but as Kant says, "to what will you fasten this point of time, if not to what is already there?" (A188/B231). It would make no sense to speak of an empty time because this could not be an object of experience, or of a *different* time because for this different time stream to have any influence on our own it must connect to it in a relation of succession and immediately annul itself. Thus, if the arising is tied to what existed beforehand and that endures up to the arising itself, then "this something that arises was only a determination of what, as the permanent, was beforehand" (A188/B231). The event can thus only be regarded as evolution and not creation from nothing.

The transcendental use of the productive imagination is primarily fleshed out in the short, inscrutable chapter of the first *Critique* devoted to the schematism, which Heidegger considers to be the central pivot of the entire work. It is here that the receptive-spontaneous character of the imagination and its position as the central point of unity between sensibility and understanding is given its clearest expression.

The schematism deals with the decidedly difficult problem of how the pure, nonempirical concepts of the understanding (the categories) come to be applied to particular cases in experience. In any application of a concept, Kant tells us, there must be a homogeneity between it and the sense impression, meaning the concept must already contain that in the object which is to be brought under it. In the case of empirical concepts this is relatively easy to understand. The empirical concept of a ball is contained within the formal geometrical concept of a sphere as the former is a concrete example of the latter. This homogeneity allows us to recognize a ball as being a case of the concept sphere. The homogeneity between concept and object, however, is entirely lacking with the pure concepts of the understanding. For example, there can be no concrete example of the mere concept of causality, as the relation of ground to consequent. Nothing in the concept of causality itself gives us the rule for its application to an object of experience. In Heidegger's language this problem is that of the unity of ontological knowledge: how pure sensibility is to be joined to pure understanding to form a whole. Kant's intricate solution to the problem relies on the discovery of a third element that can form a bridge between the two separate aspects of our cognition. This third element must be nonempirical and homogeneous with both the activity of the pure understanding and the receptivity of pure sensibility so that it can act as mediator between them. Such an element is named as the transcendental *schema*, which is a product of the pure imagination. The schema is the a priori restricting condition which allows the concepts of the understanding to be applied to possible objects of experience. Above, we outlined the way the empirical imagination forms an image out of the sensible manifold of intuition, but the schema is not a simple image. On the contrary, it is the "universal procedure of the imagination for providing a concept with its image" (A140/B179–80). To return to the geometrical concept of the sphere, no image could ever be equal to the concept of a sphere as such, which encompasses all possible spherical images. So the schema of sphere exists only in thought and never in experience, and it acts as the rule or guide governing the application of its concept. The same applies to an empirical concept, for no particular bird we could encounter in experience could adequately attain to the mere concept of a feathered vertebrate with wings and a beak. But this schema of the concept "bird" is not a generalization of every possible species of bird mangled together into one monstrous hybrid; in fact, it is not an image at all, but is rather a rule allowing for the image of a bird corresponding to the concept to be formed and for an actual bird to be identified as such.

When it comes to the categories, however, there can be no image at all. After all, how could we envisage forming an image of causality, for example? Since the process by which the categories are rendered applicable to objects of sensibility could only be pure, the key to their schematism must be sought in the *pure* form of sensibility: time. The transcendental schemata are thus defined as "*a priori time determinations* according to rules" (A145/B184). The only way the pure concept of substance can be given content is by the schema, which is the "permanence of the real in time" (A144/B183). Likewise, the schema of reality and negation are given as either filled or empty time, and so on. Thus are the categories restricted in their applicability as a very condition of their applicability. Without the schema, which limits their use to objects of sensibility, the categories signify nothing.

If we were to start with the two heterogeneous poles of pure apperception and pure sensibility (time), with no intermediary, there is no conceivable way of bringing them together into a unified whole. Time, by itself as pure intuition, is completely undetermined and unrelated to the categories, which in turn, taken alone are purely abstract and have no application to time. The transcendental schemata allow for time to be brought to unity and thus determined for thought, and on the side of thought it enables the application of the categories and gives them content. Unschematized concepts signify nothing, but undetermined intuition is likewise nothing for us, so in forming the schemata the imagination provides an indispensable role. It is clearly not the case that we need only intuition and thought in order to possess knowledge, for this only gives us the bare bones and fails to account for how they relate to one another. Only the schemata can complete the picture and allow for both sides to correspond, but it is a particularly slippery operation that it performs, being neither on the side of thought (the categories) nor intuition (time). As soon as it becomes one or the other it loses its function as mediator, but the only way it can be conceived at all is by assimilating it to one side or the other. It therefore vanishes as soon as we try to grasp it.

It is now apparent why the imagination plays so key a role in Heidegger's interpretation of Kant, and why the schematism is of such pivotal importance. Heidegger reads the *Critique of Pure Reason* as a philosophical groundlaying, working to establish the very possibility and foundations of metaphysical enquiry. Kant's investigation into the possibility of a priori synthetic judgements beginning with the pure forms of sensibility and understanding and going on to account for their unity is, according to Heidegger, nothing less than the possibility of ontological knowledge, which necessarily precedes empirical knowledge. In Heidegger's terms this concerns the *passing beyond* (transcending) given beings to the horizon upon which these beings can originally come to be. Thus Kant's project is violently grafted onto Heidegger's own in a virtuoso interpretation that enacts the point at which scrupulously close reading turns into radical reinvention.

Ontological Creativity

Heidegger's reading hinges on an analysis of receptivity, the true import of which is said to be completely missed by Kant's interpreters (such as the Marburg School of neo-Kantianism) who focused primarily on the role of logic in the first *Critique* and ignored the cardinal position of the transcendental aesthetic, and hence the primary status of intuition. However, before anything can impress itself upon us in intuition, Heidegger insists, there must be a precursory act of orientation, which lets the beings thus given *be*, as in, be revealed:

> In this original turning-toward, the finite creature first allows a space for play [*Spielraum*] within which something can "correspond" to it. To hold oneself in advance in such a play-space, to form it originally, is none other than the transcendence which marks all finite comportment to beings. (50)

This prior "act," which is a nonvolitional act that is as passive as it is active, must take place *before* we can be faced with any being which our intuition can receive or our understanding act upon. We must have projected in advance the horizon upon which this encounter can take place, and this horizon is nothing less than our understanding of being. This always-already disclosed / disclosing opening of being and its covering-over in the history of western philosophy is of course the driving problem behind Heidegger's entire philosophical project.

In this precursory ontological knowledge we do not direct ourselves toward *beings* because it is only through this knowledge that beings are enabled to be. It must pass beyond beings to their horizon, but from the perspective of beings this space we are directed toward can only be called a *nothing*, because it is not a being. Being "is" not and cannot *be*, otherwise it would be *a* being, an entity. In spite of this, however, it is still given to us (as the *giving* itself), and this opening must be kept "in view" (87) but cannot be intuited as such. In Heidegger's Kant interpretation this is where the transcendental object comes in, that unknowable X which acts as the "terminus of the preliminary turning-toward" (86). It cannot be perceived as an object precisely because it is the horizon of objectivity itself.

Ontological knowledge, then, is coextensive with the transcendental: "[this] a priori unified whole made up of pure intuition and pure understanding, united in advance, 'forms' the play-space for the letting-stand-against in which all beings can be encountered" (54). Because of the central place occupied by the transcendental imagination in this unity, as demonstrated by the mediating act it performs in the A-Deduction and its role in the schematism, imagination becomes the basis on which the possibility of ontological knowledge is founded, and hence the possibility of transcendence. In fact, for Heidegger, imagination is not only the unifying central faculty but is even the *root source* of sensibility and understanding, acting as the ground from which they emerge. The free, aspect-forming role of the imagination is the projection of the horizon of transcendence and this is necessarily prior to receptive intuition and spontaneous thought, allowing them to be what they are. From this it follows that *in themselves* pure intuition and pure understanding *are* imagination: its formative act makes their functions possible.

Pure intuition, as pure, does not receive something present, because it must be prior to all empirical intuition. Since its pure form, time, is nothing outside of the subject, but rather enables these beings outside the subject to be intuited, it is in fact *given* by pure intuition. It "procure[s] a look" (73) in advance of experience, hence pure receptivity *in itself* is pure spontaneity: the giving which allows that which *is given* to be received. What the mind intuits in pure intuition is its own activity, so pure intuition is pure auto-affection. However, since this intuition by itself could only ever intuit "the actual *now*, but never the *now*-sequence as such" (179), and because the "actual now" is an abstract fiction, it actually intuits nothing at all without the prior synthesis of the imagination, which presents time as pure arising and passing away. This can be confirmed by reference to the schematism: the "actual now" as opposed to the "now sequence" would be the *undetermined* intuition of time. It is only through the schemata that we are able to, as Kant notes, intuit "the *time series*, the *time content*, the *time order*, and finally the *time sum total*" (A145/B184–5).

In a similar move, pure, spontaneous understanding is shown "in itself to be pure receptivity. The rules of connection which the understanding deploys are not "grasped as something at hand 'in consciousness'" which we command at will but instead only manifest themselves in the act of their application. They are "binding," that is, they exert force upon us, "in their character as binding-together" (108), or connecting the manifold. The understanding gives this order and unity to experience in and through its very submission to it; hence it is only through *receptivity* to a self-imposed necessity that it is capable of the act of synthesis. Therefore, just as the "spontaneity" at the root of pure intuition is what enables its receptivity, here the reverse is true and spontaneity implies "placing oneself under a self-given necessity" (109). In this way pure intuition as spontaneous receptivity and pure thought as receptive spontaneity are reduced to functions of the transcendental imagination. Both are characterized as self-giving: as forming that which it is able to receive, whether through the auto-affection of time or the self-imposition of rules that bind.

However, does Heidegger's characterization of pure intuition as self-giving spontaneity not misrepresent the specific character of sensibility? If this act is originary and *prior* to receptivity would it not mean that the latter is derivative and that on a more primordial level our receptive finitude is overcome? After all, the very mark of human finitude for Kant is that, unlike the divine mind, ours is irreducibly divided into faculties of sensibility and intelligibility, so does not treating these poles as two stems from a common root and tracing this back to a site of original unity not lead to a conflation of human with divine cognition? As the German philosopher Heinrich Levy remarked in an early review of *Kant and the Problem of Metaphysics*,

> Has not finite pure intuition and finite pure understanding herewith obtained the structure of the Kantian *infinite intellectual intuition* and the *infinite intuitive understanding*—and indeed as completely as possible? Has not Heidegger thereby established even that his interpretation—and thereby his philosophy to which this interpretation leads—is more related to German Idealism than to Kantian criticism, despite the fact that he believes it moves in a contrary direction to German Idealism?[22]

The recourse to intellectual intuition in post-Kantian German Idealism, which Levy here alludes to, arose out of the attempt to resolve Kant's problematic duality of sensibility and intelligibility. The argument is that Kant himself is never able to put back together what he has separated and account for how two completely independent faculties performing entirely heterogeneous functions are able to interact with each other. At first glance, Heidegger's holism, appealing to the imagination to unify the two branches of "ontological knowledge," does indeed resemble the path taken by Maimon, Fichte, Schelling, and Hegel, but it is crucial to stress their divergence and to understand why Levy's critique misses the mark.[23] Heidegger stresses throughout the *Kantbuch* that the spontaneity of the transcendental imagination is not to be conflated with the creation of an intuitive intellect. The aspect-forming "act" of transcendence (which, once again, is not a volitional act) is, says Heidegger, *ontologically* creative but not *ontically* creative. The question, as Heidegger asks rhetorically, is whether this "bursts the finitude of transcendence asunder, or whether it does not just plant the finite 'subject' in its authentic finitude" (87). For Heidegger it is precisely *because* we are finite beings thrown into the midst of beings that we must be capable of holding open this horizon of transcendence, which is performed by the transcendental imagination. Ontological creativity, far from being an originary cry of "let there be light," is simply another way of saying that *being* is dependent upon Dasein's understanding of being but *beings* are not. Dasein furnishes to itself the horizon of intelligibility upon which beings reveal themselves to it, but those beings thus revealed remain enigmatic, subsisting outside of Dasein's power and indifferent to its knowing them. Entities exist independently of us but the possibility of their being "discovered" or "concealed" rests on Dasein. So ontological creativity *is* finitude itself, necessitated because we are not capable of intellectual intuition and do not have beings under our power. However, this finitude can never be raised to the status of an absolute, for the philosophy of finitude must itself be finite and never claim to have arrived at the "final truth" of finitude. Thus Heidegger's response to the question "why finitude" would be to say that it is premised on a misunderstanding: if we were in a position to provide an answer we would have lost the very thing that we are trying to account for and we would not be the beings capable of posing the question in the first place. The answer thus cancels out the question.

Kant is concerned to ground finitude on the steady dual foundations of intuition and understanding, and although we cannot inquire into the *origin* of these foundations we can at least thoroughly delineate their properties and know that nothing precedes them. Heidegger, however, by reducing this duality to branches of the transcendental imagination is by no means excavating a more primordial but no less steady ground. Rather, his gesture is to destabilize the ground altogether, for the imagination is not a solid foundation but an abyssal nonground: at once spontaneous and receptive, it rests on nothing other than its own performance. Heidegger accounts for the reduced role the transcendental imagination plays in the second edition of the *Critique of Pure Reason* as Kant's recoil from the abyss that he had uncovered. In regressing to this groundless ground, this does not mean that we have surpassed our finitude and thus accounted for it. The spontaneity at the source of our receptivity does not amount to a free creation of oneself, or the total mastery over our being. For as Heidegger declares in a typically elliptical remark in *Being and Time*, "Dasein is not itself the *ground of its being*, because the ground first arises from its own project, but as a self, it is the *being of its ground*."[24] In being its ground, *Dasein* understands its situation as thrown and grasps the possibilities that arise from this thrownness and takes them up as its own. It is not the *ground of its being*, however, for the very reason that it is thrown and not self-created, and does not give itself to itself.

For Heidegger, Dasein does not "have" possibilities as something like the property or possession of a self, something that the already existing *I* can *do* or *be*; rather, Dasein *is* its possibilities. The way I relate to and understand myself is as possibilities of myself: what I will be, what I can be, and what I am always in the process of being but without ever arriving at it. Thus "[Dasein] is existentially that which it is *not yet* in its potentiality of being."[25] This is what Heidegger calls *Ek-sistence*, as being ahead of oneself in projecting upon or being drawn toward one's own potentiality of being. However, these possibilities are not limitless but are conditioned by the withdrawal of other possibilities.

As finite beings we are thrown in the midst of beings that are neither of our making nor of our choosing, but the very process through which we find ourselves thus thrown is in the projection of possibilities. Only in *surpassing* the beings which surround me do I first find myself among and "absorbed" by such beings. "*Transcendence means projection of world in such a way that those beings that are surpassed also already pervade and attune that which projects.*"[26] I discover the chair in which I am sitting, for example, not through studying its properties but in using it for relaxation or work such that the chair "itself" does not even show up for me. Far from being *obscured* by familiarity, the true being of the things of everyday use is only thereby revealed. If I approached the chair objectively in a detached attitude and viewed it as a piece of design or as an anthropological artifact then I would miss what it "is" in the world to which it belongs, that is, something ready-to-hand and useful. The *world* is the sphere of significance such useful objects belong to, forming a referential chain of "what-for" and "in-order-to" functions, anchored in a "for-the-sake-of-which" that is Dasein itself. I am not just one being among others and do not exist *alongside* the chair that I sit in. In using such beings through world-projection, which is for the sake of my own possibilities of being, I surpass them, and only thus do I ground myself in the midst of beings.

In projecting upon future possibilities I discover myself as thrown, and only through thrownness am I free to project upon these possibilities of myself. I do not "first" find myself among beings and then carve out possibilities for myself, nor do we each start out from a state of ideal, unconditional freedom which is only subsequently limited by our factical surroundings as in the infinite intellectual intuition of Fichte and Schelling. In being-ahead-of-itself Dasein is coconstituted as *being-already-in* the world and *being-amongst* innerworldly beings, and this forms a composite whole rather than individual moments that jostle and restrict one another. From this it follows that freedom is constituted by a twofold *unfreedom*. Whatever I choose is marked and conditioned by those choices that I did not take and was unable to take: first through Dasein's inability to get behind its ground and second as a result of the fact that taking on possibilities of being is simultaneously the *turning away from* other possibilities.

To spell out this unfreedom more clearly, the possibilities open to me spring from the factical circumstances in which I find myself from birth and which I did not choose. This situation furnishes the condition of all of my projected possibilities though at the same time precluding other possibilities that would have been open were I to have found myself in another set of historical or geographical circumstances. So before any choice has even been made those choices are necessarily limited. Then, in every choice that I do make other openings are constitutively closed off, so every decision is shaped and constituted not only by what I *did* but by what I did *not* do: the choices that I did not take, even if these "other possibilities" were never apparent to me.

As Heidegger explains in "On the Essence of Ground," "transcendence at once exceeds and withdraws."[27] It is *excessive* because in being ahead of ourselves we are always more than what we are, but it *withdraws* because my freedom is permeated with nullity due to the retreat of other possibilities. This nullity, rather than being simply negative, "first brings those possibilities of world-projection that can 'actually' be seized upon *toward* Dasein as its world."[28] It is thus not an external limitation on our freedom but belongs *internally* to its structure, so that possibilities are only possibilities because they are always preceded and accompanied by the withdrawal of other possibilities. This is exactly why freedom, for Heidegger, is finite freedom, and *infinite possibility* amounts to *no* possibility, or *im*possibility. "God," Heidegger tells us, "has no possibilities in the sense that he might be something specific that he is not yet but could only come to be."[29]

Similarly, Sartre insists that a freedom which did not face "resistance," or "obstacles," is inconceivable and contradictory. If the "ends" or "possibles" which I project myself toward were *immediately* realized—"if it were sufficient to hope in order to obtain"—then my very subjectivity and sense of self would be at stake.[30] For if it is out of my future projections that I know and am conscious of myself, then were these projected possibles to become automatically fulfilled, "no project of myself would be possible since it would be enough to conceive of it in order to realise it. Consequently my being-for-myself would be annihilated in the indistinction of present and future."[31] So "once the distinction between the simple *wish*, the *representation* which I could choose, and the *choice* is abolished, freedom disappears too."[32]

From a cognitive science perspective, Thomas Metzinger writes that "goal-directedness functions as a source of stability" for an intentional system such as ours, while external reality "presents constant perturbations and instabilities."[33] An indefinitely extended period of stability brought about by the overcoming of all obstacles would lead to an attenuation of self-consciousness, so it is precisely in those moments of disturbance and incongruity between what we expect and what we perceive that "a conscious self emerges."[34] Exactly the same point is uttered by the narrator of Thomas Bernhard's novel *Correction*: "[the] world around us is constantly balking and hindering us and it is precisely by this constant inhibiting and hindering action that it enables us to approach our aim and finally even to reach it."[35] The a priori removal of hindrances and interruptions would leave us with nothing to accomplish and nothing to desire, echoing Hubert Dreyfus's description of an imagined "simplified culture in an earthy paradise . . . in which the members' skills mesh with the world so well that one need never do anything deliberately or entertain explicit plans and goals."[36]

The resonance for our central problem is clear, for neurotechnologies are premised upon attaining the immediate coincidence of thought and event, intention and act, possibility and actuality, thus overcoming the very "resistance" or delay constitutive of our freedom and our "being-for-self." Furthermore, this closure of the possible in its automatic realization would, for Heidegger, necessarily entail the withdrawal of the *actual* along with it. As we have already seen, the only way in which I can encounter a being (as actual) is in pressing ahead upon a future possibility, through which that being shows up *as* and *for* something. So without this delay that constitutes our finitude we would lose the very horizon of transcendence, or understanding of being, that makes up our *Dasein*.

This brings us to the third point of the citation from Žižek with which we opened this chapter, namely that closing the "gap of finitude," rather than bestowing a far greater freedom upon us, would actually *deprive* us of our freedom. It is clear why this is the case for Heidegger, the preeminent thinker of finitude, but it is yet to become clear in what respect it is also true for Kant, such that anything approaching intellectual intuition in a finite being would lead to the loss of that being's spontaneity.

From Autonomy to Automaticity

The problem of freedom, for Kant, is that of accounting for how anything like a free act can occur from out of the binding necessity of natural laws. As we saw earlier, Kant considers the consistency of the laws of nature to be the sine qua non of all experience. We can never happen upon a free cause because every temporal event succeeds a prior state upon which it is consequent and no matter how far back we regress in the series of conditions we will never discover the unconditioned. A free act cannot be accommodated within this series without either destroying it or conjoining with it and annulling itself as freedom. However, it does not follow that because the known laws of nature are necessary for human experience they are necessary absolutely. This would be to commit the cardinal error of dogmatic metaphysics, namely, of applying the concepts of the understanding beyond the bounds of experience. So while in the world of appearance all events must occur according to the law of cause and effect and no such thing as a free cause can ever be found to exist, things in themselves are not subject to this condition. Being extra-temporal, they can only be considered as the free and unconditioned ground of appearance. One and the same event viewed now under the sensuous aspect and now under the intelligible could be *both* naturally conditioned and free. As regards the event as appearance, it is conditioned by what preceded it in time, but as regards the very same event on the side of the thing in itself, it is free.[37]

What remains to be explained is how this rescues the notion of human free-will, which is after all the only reason the question of freedom becomes a problem in the first place. If freedom exists only in the noumenal dimension (of which we have no knowledge whatsoever) freedom of the will becomes a mystical power belonging to some unknowable, intelligible "self" which pulls the strings of our phenomenal self while we remain perfectly unaware of who is really in charge. Secondly, since the split into appearance and thing in itself applies to all beings, what makes a human being any different from an acorn? Is this likewise causally determined as phenomenon while free in the noumenal sphere? However, all Kant needed to show here is that freedom can logically coexist with nature and can be conceived without contradiction, for only once *transcendental* freedom has been assumed as a possibility can *practical* freedom (that is, freedom of the will) then be taken up.

The reason man differs from the rest of the natural world is that in addition to sensible receptivity he also cognizes himself through pure apperception: "viz., in actions and inner determinations that he cannot class at all with any impression of the senses. And thus he is to himself, indeed, on the one hand phenomenon, but on the other hand—viz., in regard to certain powers—a merely intelligible object" (A546/B574). Over and above the stream of sensuously given material I am also conscious of my own intellectual act of synthesis—the original unity of apperception—through which "I am not conscious of myself as I appear to myself, nor as I am in myself, but am conscious only that I am" (B157). Thus, more than mere sensible appearance but still not consciousness of a noumenal self, I think myself in pure apperception only as the *consciousness of thinking:* the awareness of the results of intellectual synthesis being products of *my* spontaneous activity. Because of this cognizance, which once again does not amount to knowledge of a purely intelligible self, we know ourselves to have not just a sensible but also an intelligible determining ground.

Therefore, in the understanding we have a faculty of the mind which is not sensibly conditioned and which is capable of guiding our actions independently of sensuous impulses. What is more, through the faculty of reason we have an ability to generate transcendental ideas which are *purely* intelligible, hence cannot be schematized and become objects for the senses, and this power has a causality of its own in determining our will. This other kind of coercion is manifested in the necessity imposed by the feeling of *ought*, which cannot be found to occur anywhere in nature. To say that a cat *ought* not to chase mice would be as nonsensical as saying that the sky *ought* to be blue, because nature knows nothing of ought, only what is, has been, or will be:

"Now this *ought* expresses a possible action whose basis is nothing but a mere concept, whereas the basis of a mere action of nature must always be an appearance" (A547–8/B575–6). So human beings are in a unique position in that alongside sensuous impulses such as hunger or tiredness we also have a purely conceptual determining influence expressed in the feeling of *ought*. If we pay regard only to the sensuous causes behind our actions, there will always be mitigating factors such as upbringing, education, economic situation, and so forth, which are outside our control and can be considered as causes of our behavior and the choices we have made. However, this does not abrogate us from responsibility for our actions because we are also guided by reason, which provides us with another set of laws, namely the moral law that elevates us above merely sensuous beings. Reason, being purely intelligible, is not subject to temporal conditions so it cannot be said to arise in time at particular junctures leading to certain effects, rather, it is the sensuously unconditioned and purely intelligible ground of our free will. All choices or actions except those governed by reason can be put down to pathological motives, so only when we act according to reason do we truly display freedom, as freedom *from* pathological drives. Nothing begins or ends in reason, it is timeless and unchanging, hence the law that reason prescribes is always the same; all that changes are the sensuous effects it has depending on the particular context.

The fact that in addition to our sensible, animal impulses we know ourselves to be subject to a different set of laws that can interrupt the chain of natural causes provides practical (but not theoretical) proof of the supersensible world of noumena. Freedom is a mere idea, which speculative reason can appeal to only problematically, whereas through practical reason freedom can be asserted as a fact: the very fact that I display freedom when I act according to the moral law. This latter manifests itself when I draft maxims of the will for myself that cannot be traced back to sensible conditions. An obvious example of such a maxim would be to never tell a lie, even if it is contrary to my own interests. However, because we can no more access the purely rational ideas governing morality through practical reason than we can through speculative reason, we cannot defer to these laws *directly* and know for certain what is the right and just thing to do in each case. Hence I govern my actions only indirectly, via the categorical imperative, rather than through immediate insight into supersensible ideas of "the good" and "justice." If I wish to judge the morality of my actions I must ask what would be the outcome if this principle (for example, never telling a lie) was to be a universal law and everyone acted accordingly. I cannot calculate it against an eternal measure of "goodness," which is a mere idea. This indirectness is a result of the fact we are sensuous (finite) *as well as* rational beings. If we did not have this limitation then we would act unfailingly in accordance with the moral law and would have no need for the categorical imperative.

In the same section of the *Critique of Judgement* that we cited above in relation to possibility and actuality, and how this division applies only because of the way our mind is structured, Kant also says this:

> Hence it is clear that it only springs from the subjective character of our practical faculty that the moral laws must be represented as commands, and the actions conformable to them as duties, and that reason expresses this necessity not as an "*is*" (an event) but as an "ought to be" (as obligation). This would not occur if reason and its causality were considered as independent of sensibility, that is, as free from the subjective condition of its application to objects in nature, and as being, consequently, a cause in an intelligible world perfectly harmonising with the moral law. For in such a world there would be no difference between obligation and act, or between a practical law as to what is possible through our agency and a theoretical law as to what we make actual. (403–4)

So if we were governed entirely by reason, in a word, *if we were not finite beings*, then our actions would be in complete accordance with the moral law and the distinction between *ought* and *is* would no longer apply. However, in such a case, where the law has been internalized to the extent that it is no longer a coercive force, would we even be speaking of a law per se? As Derrida has emphasized, no one more than Kant has stressed the indissociability of law and force; while there are laws that are in practice not enforced, "there is no law without enforceability and no applicability or enforceability of the law without force."[38] A law (in this case, of morality) that would be followed or applied *without* the aspect of force, without being experienced as (either external or internal) coercion, is not a law. Our relationship to the law, Derrida writes elsewhere, is an "interrupted" one: it commands us while remaining inaccessible to us.[39] If I were able to reach it and truly know it then its power over me as law is annulled.

The following of laws hence applies only to finite beings and not to God. In the second *Critique* this is what is named *holiness* of the will as opposed to *duty*, the former being the property of a "maximally perfect being," with the latter pertaining to "every finite rational being."[40] It is the lot of the finite being to strive after the ideal of holiness, "in an uninterrupted but infinite progression,"[41] without ever becoming equal to it. If we were to become totally equal to it then it would mean there being no possible desire or temptation at all to stray from the moral path. However, such a state of absolute autonomy would be indistinguishable from automatism. So if we were mere animals without reason our actions would be determined entirely by sensuous impulses and instincts, but if we were purely rational beings altogether free from animality we would be controlled by a higher but no less inhibiting set of "laws." It follows that we are only truly free insofar as we are *not* governed wholly by reason and do *not* have immediate access to the moral law, for knowing it would mean being unable to do otherwise. These laws are timeless and unchanging, so acting in complete accord with them would entail always acting the same way, like morally perfect machines. The paradox here, which deserves to be emphasized, is that we only display freedom when we act according to reason and the moral law, but if we acted *completely* in accordance with reason then we would *no longer be free*.

This is where we come back to Žižek, whose contention is that Kant's own identification of freedom as noumenal fundamentally misses the true implications of his own thought, failing to recognize that freedom is neither phenomenal nor noumenal, but operates in the gap in between. As we saw above, the act of "I think" (the transcendental unity of apperception), which accounts for our spontaneity, belongs neither to the world of phenomenal appearance nor to that of noumena; it is, as Žižek says, "trans-phenomenal."[42] It is an empty logical construct that is the vehicle of all empirical thought, thus "in itself it is nothing and it is known "only through the thoughts that are its predicates, and apart from them we can never have the least concept of it" (A346/B404). However, merely asserting the fact that I do not know *what* I am but only *that* I am is not enough, says Žižek, rather we must add that "*this lack of intuited content is constitutive of the I*."[43] I can only think and act as a spontaneous being so long as my "true" noumenal self (as Kant describes "this *I* or *he* or *it* (the thing) that thinks" (A346/B404)) remains inaccessible to me. Žižek maps the *I* of transcendental apperception onto his reading of the Lacanian "subject of the enunciation," and the void of this empty formal nothing is filled out with "fantasmatic 'stuff,' " which forms the self-identity that we carve out for ourselves (the "subject of the enunciated") and that can never be identical to the mere "I think," whose "notion can never be filled out with intuited experiential reality."[44] The subject, for Žižek, lacks its place in the "great chain of being,"[45] and cannot be integrated into reality. This neither/nor status of transcendental apperception is the reason why we are fundamentally "out of joint" and hence free. If mere consciousness of this thinking subject were to amount to an *intellectual intuition* of the self, "*I would thereby lose the very feature which makes me an I of pure apperception*; I would cease to be the spontaneous transcendental agent that constitutes reality."[46]

Žižek identifies the same logic at work in Kant's ethical philosophy, in a passage he is fond of quoting from the *Critique of Practical Reason* bearing the grandiose subheading "On the Wisely Commensurate Proportion of the Human Being's Cognitive Power to His Practical Vocation." Here Kant poses the question, "supposing now that nature had been compliant to our wish and had conferred on us that capacity for insight or that illumination which we would like to possess or which some perhaps even *fancy* themselves actually possessing"—namely, intellectual intuition—"what, presumably would be the consequences of this, as far as one can tell?" In such a case, "*unless our entire nature were at the same time transformed*" [my italics]:

> [Instead] of the conflict that the moral attitude now has to carry on with the inclinations, in which—after some defeats—moral fortitude of soul is yet gradually to be acquired, God and eternity, with their dreadful majesty would lie unceasingly before our eyes. . . . Transgression of the law would indeed be avoided; what is commanded would be done. However, the attitude from which actions ought to be done cannot likewise be instilled by any command, and the spur to activity is in this case immediately at hand and *external*. . . . Therefore most lawful actions would be done from fear, only a few from hope, and none at all from duty; and a moral worth of actions—on which alone, after all, the worth of the person and even that of the world hinges in the eyes of the highest wisdom—would not exist at all. The conduct of human beings, as long as their nature remained as it is, would thus be converted into a mere mechanism, where, as in a puppet show, everything would *gesticulate* well but there would still be *no life* in the figures.[47]

The difference between this passage and that which we took from the *Critique of Judgement* is that the earlier citation envisages how a perfect being of reason without sensuous limitations would act, while this passage speculates on what would happen if *we*, finite beings with our particular sensible makeup, were to be granted absolute insight into the noumenal dimension. The consequences, as we can see, would be catastrophic. The being of reason would act in total accord with the moral law because it is morally perfect, while the finite subject granted divine insight would act out of terror. The upshot in each case is that we would act according to the moral law not through choice or rectitude but because we would be incapable of doing otherwise, hence this would be no kind of freedom, nor morality, worthy of the name. True morality, for Kant, consists not only in acting according to the letter of the moral law but in doing so because it is our duty. Any "objectively" good deed that was carried out inadvertently, or solely for our own gain, would not be a moral act. So in gaining insight into the noumenal domain we would unfailingly fulfill our moral obligations but out of fear rather than duty and hence it would neither be moral nor free behavior. As Žižek writes,

> [The] inescapable conclusion is that, at the level of phenomena as well as at the noumenal level, we—humans—are "mere mechanisms" with no autonomy and no freedom: as phenomena we are not free, we are part of nature, "mere mechanisms," totally submitted to causal links, part of the nexus of causes and effects; as noumena, we are again not free, but reduced to "mere mechanisms." . . . *Our freedom persists only in a space between the phenomenal and the noumenal.* It is therefore not that Kant simply limited causality to the phenomenal domain in order to be able to assert that, at the noumenal level, we are free autonomous agents: we are free only insofar as our horizon is that of the phenomenal, insofar as the noumenal domain remains inaccessible to us.[48]

The irony here is that gaining "the insight which we would like to possess," which we think would enable us to throw off the shackles of bondage on the level of nature, could be achieved only at the cost of a higher form of enslavement. So freedom must be conceived as the *act* which arrests the causal process and consists of the effects that the use of reason has *in the world of appearance*, rather than a state belonging to a purely intelligible self. However, because it must always be possible to trace back any event in the world of appearance to an antecedent cause, this *act* of freedom is, properly speaking, undecidable. We must assume its existence, to avoid our being reduced to "mere mechanisms," and yet we can never say in good faith or with any degree of certainty that we have exercised our freedom, or acted morally, rather than according to biological impulses or inherited, habitual patterns of behavior.[49] It is thus clear that for Kant, too, in spite of his own words to the contrary, freedom is necessarily finite and stems from the fact that we are more than just a part of nature—that is, transcendent—but yet without being beyond it. We are free to the extent that we are suspended between two forms of automatism: a natural and a moral automatism, and this is precisely our finitude. So for Kant as well as for Heidegger, any seemingly desirable overcoming of finitude would lead to the loss of that very freedom that it seeks to enhance. Greater freedom always necessarily remains entangled within an economy of unfreedom.

Quentin Meillassoux and Intellectual Intuition

At this stage we would do well to pause and take stock of where our discussion has taken us. Given what has been set out so far, is there any credence to the suggestion that a technological instrument, however advanced we may conceive it to become, would have the capacity to bestow the faculty of intellectual intuition onto human beings? Refraining from the immediate temptation to simply reject this as exorbitant, it would be worth trying to envisage the experience of operating such a technology and the potential power it would confer upon us. Heidegger speaks of the "insecurity" of mastery, and the impotence inherent within potency. Whatever control we exert over beings stops short of full mastery due to their enigmatic opacity and autonomy: they are outside of our power and 'are never of our making'.[50] The ability to command and manipulate beings via thought alone would, however, surely amount to a greater degree of control than that which we have ever possessed before. Our intention alone brings about the desired act from the machine. However, we are still subject to practically all of the same limitations as before: the tool that we control with our mind cannot do any more than when it is controlled by more conventional means. We ran up against similar limits with regard to the brain-computer interface for music and how we could not expect it to channel previously untapped creative talent. There is a sense here in Žižek's claim that bypassing the body equates to overcoming finitude, and that is far from being self-evident.

Jack Gallant's cognitive imaging technology, which scans the brain and reconstructs a live stream of inner phenomenological experience, comes somewhat closer to attaining that divine faculty of intellectual intuition. Superficially at least, the prospect of "creating" an object through an act of thought alone would seem to be akin to the divine faculty, but as we have repeatedly drawn attention to, the thought that generates this object is itself structurally dependent on receptivity and finite constraints and the object brought into being by thought is subject to physical laws which we do not determine ourselves. The "objects" of the divine mind are not assembled from receptively intuited data and projected outwards into space but arise *ex nihilo* in and for the act of intuition: *standing forth in the act of letting-stand-forth* in the words of Heidegger. This "object" thus intuited (Ent-stand) is a purely intelligible entity, which is structurally obscured by the constituting act of our finite, sensuous cognition. So no matter how much psychical control over external reality we may be afforded by neurotechnologies, this limit remains in place, forever restraining us from attaining intellectual intuition in the true Kantian sense.

However, the lure of intellectual intuition retains its hold, despite these cautions. In addition to Žižek's recent appeal to the concept it has made a further unexpected return, in the context of a polemical critique of Kant, in French philosopher Quentin Meillassoux's highly influential 2006 book *After Finitude*. The target of Meillassoux's denunciation is the post-Kantian consensus which Meillassoux dubs *correlationism*. This consists in "disqualifying the claim that it is possible to consider the realms of subjectivity and objectivity independently of one another."[51] Moreover, Meillassoux contends that Kant's negative epistemological claim has since been hardened into a positive ontological thesis. So while Kant holds fast to the thought of an *in itself*, structurally separate from our knowledge, in subsequent philosophies, particularly the post-Husserlian school of phenomenology, the very idea of a reality "in itself," that is not *for* some transcendental consciousness became completely unthinkable. So just as there is no subject that is not already related to an object (the paradigmatic case here being Heidegger's account of being-in-the-world), likewise there is no possibility of thinking an object abstracted from its relation to a subject. This reciprocity of the terms, or "the correlation," is considered to be *prior* to the terms themselves and each only *is* in so far as it is for the other.

As Meillassoux writes,

> [One] could say that up until Kant, one of the principal problems of philosophy was to think substance, while ever since Kant it has consisted in trying to think the correlation. . . . [To] discover what divides rival philosophers is no longer to ask who has grasped the true nature of substantiality, but rather to ask who has grasped the more originary correlation. (6)

This correlationism, Meillassoux tells us, pervades analytic philosophy every bit as much as the post-phenomenological continental schools, manifesting itself in the guise of language in the former and consciousness in the latter. Both provide us with our only access to externality, while simultaneously keeping us at an infinite distance from it. We can gain no neutral, exterior vantage point from which to observe either ourselves or the world because we are imprisoned within interiority, but it is an interiority which is "transparent," offering us a partial perspective on the outside.

Philosophy has thus retreated into an ever-shrinking corner, denying to itself the tools with which to think what Meillassoux evocatively calls the "*great outdoors*" (7). This means that any scientific truth-claim about the nature of being cannot be confirmed or denied through reference to its intrinsic adequation to the phenomena under investigation but only via the present intersubjective agreement of the scientific community. Moreover, this correlationist circle is seemingly impossible to exit and there is a watertight set of arguments designed to halt anyone who attempts to do so, all of which follow the same lines: the insistence that there is no way for thought to know what *is* when there is no thought, since it cannot escape itself and isolate this supposedly self-subsistent reality from the thought about it. All realism is characterized as naïve, failing to grasp the originary status of the correlation. However, Meillassoux purports to have uncovered a devastatingly simple breach in the circle with what he calls the problem of the "arche-fossil," and the domain of *ancestrality* that such phenomena attest to. If a fossil in the usually understood sense of the term is a petrified impression bearing traces of ancient animal and plant life, the arche-fossil points to a period—"ancestrality"—anterior to the emergence of life itself. Such indicators allow scientists to form knowledge about events that occurred billions of years prior to the emergence of humankind. The philosophical significance of such scientific claims for Meillassoux becomes clear once we ask how the correlationist is to account for them. Here is what Kant has to say about scientific investigation into the origins of the universe: "because the world does not exist in itself at all (i.e., independently of the regressive series of my presentations), it . . . is to be met with only in the empirical regression of the series of appearances, and not at all by itself" (A505/B533). Therefore the truth of the referent in such a claim is not what is at issue, only what the claim purports to be about. For Kant we can indeed interrogate as far back as our scientific instruments allow but what we uncover in such investigation exists only in and for the "empirical regression," hence not *in itself*, prior to its being known and merely waiting to be discovered. So the scientist's claims will be valid as far as it goes, so long as he does not assert that such findings would be true independently of his knowing it (thus claiming to know what cannot be known).

Regardless of how much it ostensibly departs from Kant, all correlationism, says Meillassoux, must add variants of this same caveat to scientific knowledge. For the correlationist, the arche-fossil cannot be the "*givenness of a being anterior to givenness*," which would be nonsensical; rather, it "*gives itself* as anterior to givenness" (14). In accounting for such data, the scientist cannot *start* from the ancestral past but can only carry out a "*retrojection of the past on the basis of the present*" (16). Science does not amass objective knowledge of a world existing in itself but only the present indicators of a past which gives itself to consciousness as existing prior to consciousness and independently of it. So again, the correlationist does not question the veracity of the claim but only supplements this claim with a subtle and seemingly innocuous qualification. However, Meillassoux insists, such a supplement does not deepen the scientific statement but annuls it altogether, such that "either this statement has a realist sense, and *only* a realist sense, or it has none at all" (17). In short, if the big bang is not a model of how the universe actually originated, independently of human knowledge about this event, then it has no meaning or use for science whatsoever.[52]

The standard supposition that the correlationist would appeal to in order to make such an event thinkable would be to add that *had there been a witness* then it would have been perceived to have taken place as the scientist describes. However, the problem of ancestrality cannot be reduced to the standard Berkeleyan "if a tree falls and there is no one there to hear it . . ." cliché, because the latter case is contemporaneous with the possibility of givenness while the former, crucially, points to a time which antedates manifestation itself. As Meillassoux writes, "the ancestral does not designate an absence *in* the given, and *for* givenness, but rather an absence *of* givenness" (21). Thus, for Meillassoux, the emergence of the finite *can* be accounted for, and quite simply: it is merely an occurrence that arrived along with the evolution of a species endowed with a sufficiently advanced central nervous system, not a singular ontological schism that splits being off from itself. This, as we have seen, is exactly the problem that could not be accounted for from within the "correlation"—the coming into being of the correlation itself, or the question of "why finitude?" Finitude then, or manifestation, is not the irreducible condition of there being a world but is an "intra-worldly occurrence" (14), prior to which and alongside which, being itself persists in absolute indifference and autonomy. Such a brusque, deflationary answer to a seemingly intractable, centuries-old problem is typical of Meillassoux's refreshingly gung-ho approach to philosophical argumentation.

However, Heidegger would offer the rejoinder that Meillassoux's "solution" rests on an illegitimate conflation of the "world," as the totality of significance, with the "universe," namely the "physical world" that is the object of scientific enquiry.[53] Moreover, when Heidegger says that the world "comes not afterward but beforehand" this is to say that only on the basis of an historical set of practices and knowledge formations can the "universe" be known in the way that we presently do, which is unarguable.[54] It does not follow, however, that "something real can only be what it is in itself when and as long as Dasein exists."[55] As Heidegger notes elsewhere: "World is only, if, and as long as a Dasein exists. Nature can also be when no Dasein exists."[56] The priority of the world over the universe thus pertains not to the facts or events known by science but to the possibility of that knowledge itself.[57] In short, the fact that a finite being with an understanding of its being evolved at a certain point in the history of a certain planet does not dissolve the question concerning how to account for that understanding. Present at hand facts (the evolution of the human species) are not sufficient to explain the *world,* which is the horizon of intelligibility through which we understand such facts. That there was a time before Dasein and will be a time after Dasein is something Heidegger is perfectly willing to accept, but this fact pertains to the "universe" and has no bearing at all on the question of finitude.

As we saw above, Heidegger insists that the philosophy of finitude is subject to that which it discloses, thus it can never reach absolute knowledge or arrive at the "final truth" of finitude. For Kant this attempt runs aground when it tries to account for the conditions of sensibility and the categories. For Heidegger it is our thrownness that acts as the ultimate limit of our capabilities and the mark of our finitude. So facticity, whether in the form of the inscrutable structure of our cognition or of our having already been thrown into a world that was not of our choosing, is thus the condition of all our knowledge projects and that which acts as their insuperable limit. However, in an extraordinary feat of philosophic reasoning, the intricacies of which we need not go into here, Meillassoux turns this argument on its head. He asserts that facticity, far from being the mere limit beyond which our knowledge cannot go, constitutes positive knowledge of the absolute. The crux of the argument is that in *thinking* the non-necessity of the correlation, which is required in order to ward off absolute idealism that absolutizes the correlation itself (*a la* Hegel), the correlationist thinks a stratum of being that is not the correlate of her thinking, precisely because it pertains to her own nonbeing, and hence the nonexistence of the correlation. The absolute thus thought but unacknowledged is the "*capacity-to-be-other as such,*" the "*possible transition,* devoid of reason, of my state towards any other state whatsoever" (56). The neat reversal that Meillassoux performs here is to show that, in order to protect herself against the Hegelian, the correlationist must aver the necessity of the correlation's contingency, but in doing so she transgresses the correlationist circle at the same time as she reinforces it. So while facticity in the negative sense signifies a subjective deficiency—our inability to fully account for our own existence and that of the world around us—facticity in the positive sense projects this absence of reason outside, such that the absolute *is* the absolute absence of reason or necessity for anything to be or remain the way it is. The crucial difference between the two is that while the former maintains a disavowed belief that such a reason may exist but be outside of our faculties of knowledge, Meillassoux categorically asserts that there is *no* such reason, *and we know this.*[58] He derives from this speculative absolute a series of consequences he names *figures*, arguing that there must be a set of invariable properties that all contingent entities (anything that is) have in common. To be contingent is not to be anything whatsoever but is to have a determinate set of conditions, such as adhering to the principle of noncontradiction. From this he concludes that "*whatever is mathematically conceivable is absolutely possible*" (117).

Meillassoux's ultimate purpose in all of this is to secure the absolute reach of mathematics, such that we can say without hesitation that the properties of a thing that can be described mathematically are properties belonging to that thing *in itself,* having no reference to thought. This he claims to have achieved through establishing thought's access to absolute contingency. However, the case for the absolutization of mathematics is not conclusively made—it is asserted rather than demonstrated[59]—and nor, as we shall see, is the problem of ancestrality ever satisfactorily reconciled with the "principle of factiality" (the nonfactical, that is, necessary, nature of facticity.) Its achievement, nonetheless, is that it does convincingly and exhilaratingly make the case for philosophy to regain its speculative scope by rediscovering thought's relation to the absolute.

However, let us interrogate this relation a little closer. If through this knowledge of the necessity of facticity we gain access to a purely intelligible absolute, just what kind of knowledge is it? Clearly it cannot be obtained through sensible intuition because according to Meillassoux it is our senses that impose upon us the belief in the necessity of natural laws, through habit and superstition, while nothing supports such an inference from a rational point of view, as Hume demonstrated. It is here that Meillassoux, somewhat surprisingly, makes his appeal to the Kantian concept of intellectual intuition:

> [We] discover in our grasp of facticity the veritable *intellectual intuition* of the absolute. "Intuition," because it is actually in what is that we discover a contingency with no limit other than itself; "intellectual" because this contingency is neither visible nor perceptible in things and only thought is capable of accessing it, just as it accesses the chaos that underlies the apparent continuity of phenomena. (82)

While evidently not being employed in a strictly Kantian sense, because here thought accesses absolute being while by no means *creating* it, this is nothing less than *noumenal insight,* as described by Kant in the passage we quoted above. However, in a move precisely contrary to Kant's, such a knowledge does not impress upon us the terrifying divine majesty of all things but rather the complete *absence* of order or necessity, whether divine or otherwise. Knowledge is thus radically divorced from the senses, indeed on this point is shown to be in direct conflict with the senses.

We are entitled to ask, however, as Kant did in the letter to Herz, how we are to guarantee the necessary reference of this intellectual intuition to the nature of being itself. The two alternatives Kant presents us with surely still apply: either the innately logico-mathematical essence of being gives itself to thought, or thought projects its logical intuition into being. However, neither option is available to Meillassoux, since taking the former would commit him to an untenable Pythagorean ontology, which he has already explicitly ruled out, and the latter, as we know, is intellectual intuition in the Kantian sense. This is why for Kant intellectual intuition could never be *receptive,* because it is impossible to conceive of how its object would be transmitted to thought. The very bedrock of Meillassoux's enterprise is the thesis that mathematics allows us access to a reality independent of thought, but if this is merely thought's projection onto reality then we are not yet free of the correlationist circle. Ray Brassier makes broadly the same point, writing that if reality is "neither inherently mathematical nor *necessarily* intelligible," why should we assume that being is susceptible to intellectual intuition?[60] If this reference is itself also intuited intellectually then too much is ceded to thought and correlationism creeps back in.

In the face of Brassier's criticisms Meillassoux modified his position, opting instead to employ the oxymoronic term "dianoetic intuition," meaning "the essential intertwining of a simple intuition and of a discursivity, a demonstration—both being entailed by the access to factuality."[61] As he goes on to explain, if in order to break out of the correlationist circle we were to merely posit an autonomous real axiomatically (as Francois Laruelle does according to Meillassoux), the correlationist will always have the rejoinder that this ostensibly autonomous real is still *posited* by thought. The only remaining strategy, says Meillassoux, is the one taken by *After Finitude*, namely to start from within the circle of correlation and demonstrate how, in order to maintain its consistency, it must itself appeal to an absolute—facticity:

> Hence, the only way to the Real, according to me, is through a proof, a *demonstration*: a demonstration unveils that facticity is not an *ignorance* of the hidden reasons of all things but a *knowledge* of the absolute contingency of all things. The simple intuition of facticity is transmuted by a *dianoia*, by a demonstration, into an intuition of a radical exteriority. . . . We have a *nous* unveiled by a *dianoia*, an intuition unveiled by a demonstration. This is why I called it an intellectual intuition: not, of course, because it is an intuition which creates its object, as Kant defined it, but because it is an intuition discovered by reasoning.[62]

An immediate intuition could never give us access to the real, because as the correlationist would remind us, we only ever intuit phenomena which are constituted by that act of intuition. But a simple logical positing of the real from which we then derive conclusions will not satisfy the correlationist either. So intellectual (or dianoetic) intuition in Meillassoux's sense is not an immediate revelation but is the logical explication of a prior intuition. Through rational demonstration this intuition (of facticity) is shown to be not what we thought it was. What had seemed to be the insurmountable limit to thought and the essence of finitude is, through intellectual/dianoetic intuition, revealed to be the key to the very overcoming of finitude.

However, as Alberto Toscano has incisively argued, in relying on purely logical reasoning to secure his "primary absolute"—the necessity of facticity—Meillassoux sacrifices the secondary absolute that forms the basis of his critique of correlationism, namely the arche-fossil or the natural kinds that are radically exterior to consciousness. So we have a materialist absolute (uncorrelated matter) and a logical absolute (necessary contingency) and no immediately satisfactory way of unifying them, for while one relies upon an irreducible gap between thought and being, the other is premised on their continuity. This is why, for Toscano, speculation and materialism are incompatible, for by "trying to maintain the speculative sovereignty of philosophical reason" and purporting to draw "ontological conclusions from logical intuitions,"[63] Meillassoux relinquishes his claim to the materialist maxim that being is not reducible to our capacity to think it. Toscano's critique is, at base, a restatement of Kant's refutation of the ontological proof of God, the latter claiming to infer the existence of an entity through pure reason alone, or, draw "ontological conclusions from logical intuitions." Indeed it is quite possible to accept the basis of Meillassoux's critical broadside against the correlationist, namely that she commits pragmatic contradiction by appealing to an absolute in order to disqualify absolutism, and also his rejection of the principle of sufficient reason, while refusing the ontological consequences that he develops from it. While contingency may indeed be necessary for the correlationist to prohibit the absolutizing of the correlation itself, this does not of necessity entail its absolute status beyond this narrow philosophical context. In other words, why is this *necessarily* anything more than an inconsistency within a philosophical position?

This brings us back to Ray Brassier's important objection: unless being is inherently logico-mathematical how can we guarantee that it is exhaustively available to intellectual (dianoetic) intuition? Meillassoux's previously cited response does not quite scratch this itch. If the correlationist will not be silenced by the Laruellian approach (the axiomatic positing of the real), it is not clear that Meillassoux's alternative will satisfy them either, by falling foul of Kant's critique of the ontological argument. The two absolutes identified in *After Finitude* correspond to the Kantian opposition between receptive and productive intuition, which, as we saw via Heidegger, structurally obscure each other. The materialist absolute maintains the gap between thought and being by not allowing the latter to be reduced to the former; the logical absolute then overcomes that gap by asserting the contiguity of logical reasoning with being. The problem of intellectual intuition still haunts Meillassoux's enterprise, despite the terminological shift. The question is whether we can square the circle and overcome the gap between thought and being without precluding the possibility of an extra-logical reality that is excessive to thought.

3 UNUS MUNDUS

After Freud published *The Future of an Illusion* in 1927, his short, polemical tract which treats of the psychological origins of religion, the renowned French dramatist and novelist Romain Rolland wrote to him with a response. In the letter he praised the book but also expressed his dissatisfaction at the absence from the discussion of what he considers to be the true source of the religious experience. This, as Freud recounts in *Civilization and its Discontents*, "consists in a peculiar feeling, which he [Rolland] himself is never without, which he finds confirmed by many others, and which he would like to call a sensation of 'eternity,' a feeling as of something limitless, unbounded—as it were, 'oceanic.' "[1] According to Rolland this *oceanic* feeling is the source of all religiosity. It is the sense that there is a resource within us that escapes the finite, embodied limits of subjectivity. This experience is exploited in various ways by the different religions, whether it be the Kingdom of God in Christianity or the sense of oneness with the universe expounded by new age spiritualism, and so on. Freud, the great demystifier, traces the origin of this affect to the all-encompassing ego of the newborn infant who as yet does not distinguish internal from external, only gradually learning to do so in the face of a lack. For while some urges may be satisfied at any time and some sources of excitation can be felt at any moment, others—"among them what he desires most of all, his mother's breast"[2]—escape him and are not under his immediate control. Desires such as these can only be brought to satisfaction by crying and screaming for attention. It is through this experience that the child first "[sets] over against the ego an 'object,' in the form of something which exists 'outside' and which is only forced to reappear by a special action."[3] The "object," discovered as not being a part of the child's own ego, thus makes its first appearance in the upsurge of desire and the experience of a lack: a need, whose satisfaction is not in the child's power. Crying is the initial and most basic technique of reaching out into the external world to bring about this satisfaction that cannot be regulated internally.

So while "originally the ego includes everything, later it separates off an external world from itself," leaving us with a "shrunken residue of a much more inclusive—indeed an all-embracing—feeling which corresponded to a more intimate bond between the ego and the world about it."[4] The oceanic is thus the spectral persistence of this lost unity, imbuing those who succumb to it with the sense that what they now perceive to be their egoic self is merely the remainder of something more expansive, to which most religious doctrines promise a return after we die.

Auto-Satisfaction, or How to Bypass Reality

This narrative of the subject's individuation through its decisive severance from the external world is fundamental to Freudian psychoanalysis. Freud famously proposes as a useful fiction the hypothesis that our highly advanced psychical apparatus developed out of the primitive need to minimize excitation (this is the "principle of constancy" or the "Nirvana principle")[5] and that the earliest stage of this apparatus had as its only function to "[keep] itself so far as possible free from stimuli; consequently its first structure followed the plan of a reflex apparatus, so that any sensory excitation impinging upon it could be promptly discharged along a motor path."[6] Subsequently this primitive organism finds that the exigencies of life disturb and upset the success of this self-contained system, for needs such as hunger cause an excitation that cannot be soothed via purely internal means. The system can only be rebalanced if an "experience of satisfaction"[7] is felt, which suspends the disturbance. The memory trace left by this experience of satisfaction will then indelibly be associated with future occurrences of the same imbalance and the organism will seek to reenact the conditions that led to this feeling of fulfillment. In other words, as soon as the need arises again the organism tries to satisfy it by means that it has previously found successful. As Freud writes,

> An impulse of this kind is what we call a wish; the reappearance of the perception is the fulfilment of the wish; and the shortest path to the fulfilment of the wish is a path leading direct from the excitation produced by the need to a

> complete cathexis of the perception. Nothing prevents us from assuming that there was a primitive state of the psychical apparatus in which this path was actually traversed, that is, in which wishing ended in hallucinating.[8]

In this primitive, self-sufficient state that Freud describes, when a biological need forced itself upon the organism satisfaction was found in the hallucinatory recall of a previous perception in which this need was satisfied. No sooner has the need or wish arisen than it is perceived to be satiated. This structure, Freud suggests, persists in dreams, where unconscious wishes are represented in the perceptual system as fulfilled, albeit in a disguised form. However, in waking life, "bitter experience" will have necessitated the evolution of this inefficient hallucinatory course of fulfillment into "a more expedient secondary one."[9] Since the hallucinatory satisfaction of a need such as hunger does not actually satisfy the biological need it developed in response to, it was found necessary to "seek out other paths which lead eventually to the desired perceptual identity being established from the direction of the external world."[10] Thus while this early stage of development is dominated entirely by the pleasure principle its place is subsequently usurped by the reality principle, which constitutes a supplementary system of deferrals on the way to the original satisfaction. External reality will not immediately satisfy our every need so we must learn to adopt a degree of pragmatism in the pursuit of our goals. Central to this is what Freud calls "reality-testing": assessing whether an internal presentation corresponds with external reality and if not, taking the necessary steps to either bring about its existence or modify our inner sense accordingly. This is where technology enters the picture, with the technical system compensating for the shortcomings of the biological system by making the external world more accommodating to our needs and our pleasure goals, easing and hastening satisfaction. All technological advances are, at base, designed to make this process easier still. However, with the neurotechnologies which bring about the automatic and instantaneous realization of the user's intention we are now facing what can only be described as the final stage of this progression, which may amount to a full return to the primitive state of autogratification.

The future heralded by the "Bluetooth in your head that translates your thoughts into actions" by circumventing the motor system would be one in which the pleasure principle has been fully reinstated to its preeminent position at the expense of the resistance of external reality. In an early development indicative of the potential mass-market roll out of neurotechnologies, the UK-based design and digital innovation company This Place developed an accessory for Google's (subsequently discontinued) wearable technology device Google Glass that allowed thought controlled navigation through its features, which included taking photographs and browsing the Internet. Named MindRDR, it positions itself as inaugurating "a new generation of mind-controlled wearables," giving us a key insight into the future consumer applications of neurotechnology. It uses the MindWave mobile EEG headset manufactured by consumer electronics company NeuroSky, which we discussed briefly in the Introduction, and its primary selling point is that users can generate "content" and upload it to social media platforms all "using the power of your mind." So depending on the sensitivity of the scanner and the speed of the processor, all of this can occur *at the moment the intention is formed* and thus before the thought has even been fully articulated to its bearer.

In a clinical setting, the application of BCI technology to domestic environments has been successfully trialed, signaling the possibility of a new generation of "smart homes" in which opening and closing doors, activating lights, answering the telephone, and operating electronic entertainment appliances are all controlled by the mind.[11] The Santa Lucia Foundation and Hospital in Rome is one of the key research centers developing domestic applications for neurotechnologies. One example is a BCI using a selective attention task similar to those outlined in our Introduction and in chapter 1. It presents the user with a visual interface displaying a cascading menu of icons representing domestic appliances which flash when users are concentrating their attention on them. When the intended action has been recognized by the BCI a command is sent to the relevant device to either enable or disable its activity.[12]

Already, many of our household appliances, from the thermostat and lighting to telecommunications and entertainment systems can be integrated wirelessly into one networked, centrally controlled system (the famous Internet of things). With the rate at which neurotechnology is advancing it is entirely plausible to imagine that at some stage within the next few decades, instead of a remote control operated by hand (or, increasingly, voice command), this will all be activated merely by thought. The television would be correlated to the user's attention-span, instantly switching channels as soon as his interest dips; a query pops into his mind and a computer promptly finds it on a search engine; he thinks of a song and it downloads from the Internet and starts playing on the stereo. However, this automatic realization of even the most fleeting desire could lead only to a state of near-total indolence, akin to the effects of the eponymous film cartridge in David Foster-Wallace's novel *Infinite Jest*, which is so lethally entertaining that whoever views it becomes instantly reduced to a catatonic state, having lost all interest in anything outside of the film.

Freud, anticipating Karl Popper, writes that thinking is "an experimental kind of acting."[13] In the metapsychological system the ego interposes itself between the desires of the id and access to motility. By these means "it has dethroned the pleasure principle which dominates the course of events in the id without any restriction."[14] Here once again we see the significance of the interval separating intention from deed, which neurotechnologies bypass. Instead of having to adjust our goal of attaining pleasure in accordance with our surroundings we can instantaneously reorganize our surroundings in accordance with our pleasure goals. Unconscious repressed processes, says Freud, "disregard reality-testing; they equate reality of thought with external actuality, and wishes with their fulfilment—with the event—just as happens automatically under the ancient dominance of the pleasure principle."[15] Waking life would follow the pattern of dreaming in which wishes are immediately fulfilled as soon as they arise, although not merely in hallucinatory form. Furthermore, given the frequency with which obscene or unsavory thoughts and desires are prone to arise unbidden into one's mind at any given moment we may find ourselves compelled to establish some form of filter such as parents of young children put in place on the Internet to prevent such momentary thoughts from gaining an outlet. This would be a technological outsourcing of our self-restraint required to counteract its technological short-circuiting, akin to the "I don't want to see this" option on Facebook's automatically generated memories.

The exigencies of reality testing, and of modifying and subordinating our instincts towards satisfaction in accord with objective reality would be radically reduced by the "Bluetooth in your head that translates your thoughts into actions." What we wish for or crave would find immediate satisfaction with no delay and no effort expended. So if external reality is that which moderates and constrains the excesses of the pleasure principle, it seems at first glance that the use of advanced neurotechnologies in the home would amount to the immediate self-perpetuating presence of untamed pleasure. However, this would be a simplification, conferring a provisional, secondary status onto the reality principle and overlooking the extent to which this economic detour of reality structures and constitutes the experience of pleasure.[16] As Derrida noted, "[pure] pleasure and pure reality are ideal limits, which is as much as to say fictions," each deriving from a common "necessarily impure" root, which is the differing and deferring movement that Derrida calls *différance*.[17] It is a truism that pleasure is often found more in the postponement of satisfaction than in the satisfaction itself and that satisfaction in fact frequently leaves us *un*satisfied and deflated. The instant gratification of any and every wish would lead rather to extreme lethargy than to some orgiastic overdose of enjoyment. As Freud writes, "[when] any situation that is desired by the pleasure principle is prolonged, it only produces a feeling of mild contentment."[18] The injunction to attain to a state of "pure pleasure" commanded by the pleasure principle, then, is one that is impossible to fulfill but equally impossible to take leave of. So everyone takes his or her own path toward this limit but none ever reaches the final destination: "It is a question of how much real satisfaction [one] can expect to get from the external world, how far he is led to make himself independent of it, and, finally, how much strength he feels he has for altering the world to suit his wishes."[19] The upshot is that in bypassing the interruption that the reality principle constitutes we bypass pleasure as well.

Not only would individuals reticulated within this autogratification network no longer be able to obtain pleasure, but they would also have no experience of an event. When the field of the possible, of what-may-come, is saturated in advance by need or by desire—in short, by the subject's own projection onto the future—there would be no opening for the unexpected to occur, or for anything truly to *happen* to the subject, unless it was on the order of a disaster (such as death, or a power failure). And where there is no delay between sensory excitation and discharge but instead a relation of pure immediacy there would be no notion even of temporal succession: past and future would merge into an indistinct plateau, "experienced" by a wholly passive, pacified, and desubjectivized organism. Such a development would bear out Freud's gambit in *Beyond the Pleasure Principle* that all techno-cultural advances are essentially regressive, constituting a "new path" to an "ancient goal": the goal of inertia.[20] This echoes our previous chapter's discussion of Heidegger and Sartre, where the removal of all resistance or delay between the aim that one projects oneself toward and its eventual obtainment was shown to be the ruin of the self (since I am my still-to-be-lived possibilities). Thus the exorbitant claims promulgated by techno-utopianists about the unimagined state of eternal happiness that awaits us if we are to fully embrace the opportunities for technological enhancement should be regarded with more than a little skepticism.[21]

Neurotechnologies represent a hyperbolic amplification of what Paul Virilio nicknamed the *citizen-terminal*: "soon to be decked out with interactive prostheses based on the pathological model of the 'spastic,' wired to control his/her domestic environment without having physically to stir: the catastrophic figure of an individual who has lost the ability for immediate intervention along with natural motricity."[22] Rather than bestowing upon us an increased ability to interact with our surroundings—a greater power of mastery or manipulation—such technological devices would, following Virilio, deprive us even more of our powers of active engagement by delegating activity to the machines, leaving us ever more estranged from our environment. However, neurotechnologies raise the stakes significantly beyond the interactive media that Virilio bemoans, which are all cases of hand-operated automation—remote controlled televisions, lights operated by clapping, and so on. Such developments undoubtedly result in a reduction in the degree of active participation but they are nevertheless still an *activity*. After all, is there really a fundamental difference between whether I operate the television from here or there, or turn on the lights by flicking a switch or clapping my hands? With brain-controlled interactive technology even that minimal externality has been superseded. Once again, as Žižek commented, " [even] Steven Hawking's proverbial little finger—the minimal link between his mind and outside reality, the only part of his paralysed body that Hawking can move—will thus no longer be necessary: with my mind I can directly cause objects to move."[23] Between operating the television by walking across the room to flick a switch and doing so remotely from where I am seated there is merely a difference of degree. However, in the passage from operating it manually to simply *intending* to commit such an action a limit has been reached.

Here the relinquishing of activity to machine assistance takes place at the very forming of the thought or intention itself, before we have even stirred. So as well as a diminution in our active involvement with the world this risks an ever-greater standardization, automation, and homogenization of thought itself. Suffice it only to recall clumsy auto-correct functions on search engines and text messaging software, erroneously altering one's messages, violently interposing itself between the writer and the written and forcing us to say what it expects us to say rather than what we actually want to say. It is no stretch of the imagination to foresee predictive algorithms completing an intention before it has been fully formed by consciousness in much the same way as Google predicts what one wants to type from the very first letter. These predictions would be based on our previous choices, as well as those of other users, thus reducing thought and action to mere repetition and habit. Nor is it especially outlandish to expect the outcome of these predictions to be shaped by commercial sponsorship (as of course Google predictions are), where our thoughts and desires are directed toward consumer products. Such automation would further entrench to a truly unprecedented degree what Bernard Stiegler terms *proletarianization*, as the outsourcing of knowledge (in the sense of theoretical knowledge as well as *savoir-vivre* and *savoir-faire*) to machines and the concomitant "disapprenticeship" of the subject leading to our increasing helplessness.[24] If this technology operates on a preconscious level then how are users to distinguish between their own thought and an automatically generated one? Here again we see this greater freedom (from the constraints of the body and of externality in general) being indistinguishable from greater bondage.

As noted above, a technology that enables the same cognitive function that thinks of an action to *simultaneously produce that action*, or one that instantly materializes our thoughts or phenomenological experiences, would significantly compromise our ability to perform reality testing. As Freud describes it,

> The other sort of decision made by the function of judgement—as to the real existence of something of which there is a presentation (reality-testing)—is a concern of the definitive reality-ego, which develops out of the initial pleasure-ego. It is now no longer a question of whether what has been perceived (a thing) shall be taken into the ego or not, but of whether something which is in the ego as a presentation can be rediscovered in perception (reality) as well. It is, we see, once more a question of *external* and *internal*. What is unreal, merely a presentation and subjective, is only internal; what is real is also there outside.[25]

Evidently these distinctions between what is *real* as opposed to *unreal*, *subjective* as opposed to *objective*, and *internal* as opposed to *external*, are now more than ever problematized and rendered uncertain. This is the same problem as that addressed in the previous chapter under the rubric of possibility (corresponding to thought) and actuality (corresponding to material existence), although there we were concerned with whether the blurring of this distinction constituted an overcoming of finitude; here it is a question of the impact it would have on our psyche and our experience of reality. After all, the immediate fulfillment of wishes is one of the chief examples Freud gives of the feeling of *das Unheimliche*, illustrated by the tale of The Ring of Polycrates. As legend has it, the King of Egypt becomes horrified by his ally, Polycrates, "because he sees that his friend's every wish is at once fulfilled, his every care promptly removed by kindly fate. His host has become 'uncanny' to him."[26] In everyday experience, when an unlikely state of affairs that we wish or hope for actually comes about, it is not uncommon to be left with a sense of uneasiness rather than pleasure, as if strange, sinister forces are at work. The reason for this, Freud tells us, is that it recalls to us our childhood belief, since repudiated, in "the omnipotence of thoughts."[27] The "unrestricted narcissism"[28] of this early stage of development, in which the reach of our mental processes is greatly overvalued, is subsequently displaced by reality. However, once again it appears that the increased potency granted to us by neurotechnologies, rather than progressing the human race to some advanced stage of posthuman development, would actually lead to a form of regression, in this case to a state of uninhibited primal narcissism.

Furthermore, what bearing would this have on the texture of our reality when "[our] whole relation to the external world, to reality, depends on our ability to distinguish [perceptions from ideas]?"[29] A dream is described by Freud as being, among other things, "an externalisation of an internal process,"[30] so neurotechnologies may put at risk our ability to distinguish waking reality from dreaming, or hallucination. This would lead to a generalized state of *amentia*, a condition characterized by Freud as the withdrawal from reality in the face of some loss or bereavement which the ego must deny since it finds it insupportable. "With this turning away from reality, reality-testing is got rid of, the (unrepressed, completely conscious) wishful phantasies are able to press forward into the system, and they are there regarded as a better reality."[31] However, just as a reversion to the state entirely in thrall to the pleasure principle paradoxically spells the *end* of pleasure, so if the psychical floodgates are opened not only to such "*unrepressed, completely conscious*" intentional fantasies as we would knowingly wish for, but to *unconscious* ones as well, those that we are not even aware that we hold, we may find this "better reality" becoming more like a nightmare.

The Id-Machine

In Freud's *Fragment of a Case of Hysteria*, his young patient Dora is revealed to harbor a powerful unconscious erotic attachment to her father's friend Herr K., but on the occasions when he actually makes sexual advances toward her she reacts violently and runs away. So why, Freud asks, would Dora fiercely reject that which she longs for the most? Why would she fear her deepest wish coming true? Freud's answer is that, as with all neurotics, Dora is "dominated by the opposition between reality and phantasy. If what [neurotics] long for the most intensely in their phantasies is presented to them in reality they none the less flee from it."[32] This, as Freud says frequently, is a general condition, for what goes for the neurotic goes also for the healthy psyche. Indeed, "we are *all* ill,"[33] it is just a matter of degrees of intensity. This is why at the climax to Andrei Tarkovsky's 1979 film *Stalker* none of the protagonists can bring themselves to enter the room with miraculous wish-realizing properties that they had risked their lives to reach. What they ultimately came to understand is that, in the words of the character named only Writer, "not just any wish comes true here, but only your innermost wish. Not what you would holler at the top of your voice. Coming true here is only what's in line with your nature, with your essence, of which you know nothing. But it's there, in you, directing you all your life." Standing on the threshold of gaining everything he ever wanted, Writer is afraid of what will be revealed to him upon entering: what he really desires without knowing it, thus bringing him face to face with his innermost truth.[34]

So that which we really desire is necessarily encrypted: concealed from us because its exhibition would disturb our psychical consistency. For Freud, our reality is necessarily supplemented with what he calls "auxiliary constructions,"[35] that are fantasmatic formations becoming so woven into the fabric of reality that they are inextricable from it. In Lacanian terms, our access to reality is necessarily mediated by fantasy but we never have conscious access to those fantasies for that very reason: doing so would tear the fabric of that reality we have built for ourselves. As a result, a considerable degree of blindness to our own actions, motives, and desires is indispensable to our ability to go about our daily lives and social interactions. However, we now live in increasingly intimate contact with digital media whose precise role is to disclose subperceptual facets of our own behavior to which we otherwise would have no access. These media include everyday devices such as smart phones with GPS tracking, wearables with built in microsensors harvesting and storing biometric data, as well as more specialized technology such as the MindWave mobile headset scanning and interpreting brain activity. While the majority of this data may be perfectly innocuous, revealing sleep cycles and patterns of movement, which are used to adjust lifestyle routines, it also carries the potential to bring to light aspects of our own experience we would find harder to integrate into our self-understanding.

Technologies such as these are the focus of Mark B. N. Hansen's recent book *Feed-Forward*. For Hansen, the significance of what he terms "twenty-first century media" (including those just touched upon) resides in the fact that, rather than mediating or extending human sense perception or cognitive faculties, they affect us at a subperceptual level, operating at "microtemporal" scales bearing no necessary or immediate relation to human conscious awareness. This stands in clear contrast to the classical function of technology, described by Freud as "auxiliary organs" designed on the model of those organs that they augment, such as telescopic or microscopic lenses extending our visual perception, or writing and photography as prostheses for memory.[36] Twenty-first century media, on the other hand, do not supervene on existing capabilities but "interface human experience with new domains to which it lacks direct, perceptual access."[37] A key example is deeply ominous U.S. military research into "operational neuroscience," one strand of which is aimed at "leveraging human perception" by harnessing the brain activity of experts trained to recognize potential areas of interest in satellite images.[38] The sensitivity and flexibility of the human brain far outstrips the most advanced machine vision technologies, however the huge volumes of data collected make it too labor and time intensive for human analysts to manually examine it all. Military scientists are working on a solution that would utilize the visual responses in the brain *as they happen* rather than waiting for the analysts themselves to compile their reports. This is accomplished by scanning the brains of the analysts with EEG machines while they view a rapid sequence of images on screen. The spikes in the EEG curve signaling potential significance are visible well before the information has been consciously registered so the technology is able to bypass the conscious experience of the human perceiver entirely and extract the desired intelligence "directly" from their brains. More sinister still is the HandShake technology we touched upon in the Introduction, which also detects signs of image recognition in the brain while viewing a sequence of images but this time with the purpose of revealing potential enemy operatives who have been betrayed by their brain signals.

For Hansen the issue here is primarily one of speed: conscious awareness is far slower and less fine-grained than our real-time bodily responses to stimuli. Consciousness can only be made aware post factum of what has been "experienced" in its absence. The process through which we become consciously aware of such data is the "feed-forward of Hansen's title, amounting to the "deceleration of a microtemporal phenomenon . . . so that it can be presentified, after the fact, to a consciousness that remains, constitutively, too slow to grasp it directly."[39]

A further example discussed by Hansen is a wearable digital device called a "sociometer," developed by MIT researchers. The sociometer uses sensors to harvest data on implicit behavioral patterns in business situations such as salary negotiations and job interviews, the purpose being to allow companies to model social interactions and learn more about the unconscious factors governing decision-making. The data collected includes nonlinguistic social signals such as tone of voice, body language, physical location, and proximity to interlocutors, which all occurs below the threshold of consciousness. Revelation of such a wealth of information will almost certainly present us with a picture of our behavior that contrasts markedly with our own experience. Thus my conscious impression of my own lived experience becomes altered in light of external data presented to me about my behavior. This digital expansion of our experiential space is distributed across two incommensurate time-scales which can only be mediated by the operation of "feed-forward": a belated, technological reappropriation by consciousness of an *experience* that cannot be *experienced* by a subject "*at the moment of its occurrence*."[40]

The broader ontological significance of twenty-first century media, on Hansen's account, is that they index a fully autonomous domain of "worldly sensibility" that exceeds and precedes perceptual consciousness and that can neither be reduced to it nor subordinated to it in terms of ontological priority. The political charge of Hansen's book lies in the way that it outlines strategies of resistance against the capture of such worldly sensibility by global corporations such as Facebook and Google who view human subjects as little more than data farms, gathering masses of information on our likes and dislikes to better predict and manipulate our future behavior. Like the U.S. military's "operational neuroscience," this "operationalization of our desire"[41] bypasses our conscious participation altogether, thereby reducing the human agent to what Heidegger calls "standing-reserve": a mere resource subjected to violent machinic extraction.[42] The task, as Hansen sees it, is not to resist twenty-first century media technologies but to mobilize these tools toward building a better future for ourselves. By feeding-forward this data of preconscious sensibility into our conscious lives (according to the complex temporal structure briefly sketched out above) we can utilize twenty-first century media to expand and reorient human experience and predictively shape our future behavior.

This is not the place to delve any deeper into Hansen's rich and original project and its revisionary rereading of Whitehead's process philosophy. However, one element that is missing from this account, which from my own point of view is absolutely crucial, is the psychical impact of this data-mining of our lived experience. Indeed, for a book which is concerned with the marginalizing of conscious experience and its embeddedness within an expanded, general sensibility, the complete absence of psychoanalysis from the picture is particularly marked. Let us take a very obvious example to drive home what is at stake here. Supposing we were to learn from the biometric data picked up by the sociometer, or a similar device, that our behavior around certain social or ethnic groups betrayed discomfort, fear, or even aggression. Given what we know from our brief acquaintance with this technology it would be relatively easy to analyze the information to these ends. In fact studies into the neural correlates of implicit prejudice have discovered "microtemporal" behavioral cues, such as startle eyeblink responses, indicating prejudice in white participants to black faces that occur in the absence of conscious awareness.[43] Social psychology studies have, since the late 1990s, frequently used the implicit association test (IAT) as an indicator of implicit racial prejudice and stereotyping. IATs measure strength of associations when classifying negative or positive attributes alongside a related pair of concepts. For instance, a variant of the test will ask the participant to categorize certain words as positive or negative by clicking specific keys on opposing sides of the keyboard, and the software measures the speed and accuracy of responses when positive attributes have been paired with images of white faces and negative words with black faces and vice versa.[44] If, for example, the respondent is quicker to categorize negative words when they have been paired with images of black faces this can be an indication of unconscious bias. The results regularly disclose a discrepancy between participants' explicit reporting of their attitudes toward race and their implicit preferences revealed by testing, so once again our conscious experience is undercut by external information harvested from our behavioral responses, revealing unconscious dispositions that may be completely at odds with our stated principles and how we understand ourselves.[45]

The question is how we are to categorize these latent, unconscious attitudes and implicit prejudices that we harbor and which influence our behavior even against our better judgement and conscious principles. They could not, of course, be termed objective as they tell us nothing at all about the object known outside of the knowing subject. However, nor could they in any meaningful sense be classified as *subjective*, as in how the object appears to the subject independently of how it is in reality, since the whole point is that it does not appear to the conscious subject this way and that is precisely why their announcement to consciousness represents such an unwelcome disturbance. So neither objective nor subjective, these attitudes instead belong to "that bizarre category" that Žižek, borrowing a phrase from Daniel Dennett, calls "objectively subjective":

> When, for example, we claim that someone who is consciously well-disposed towards Jews nonetheless harbours profound anti-Semitic prejudices of which he is not consciously aware, are we not claiming that (insofar as these prejudices do not reflect the way Jews really are, but the way they appear to him) he is not aware of how Jews really seem to him? Or, to put the same paradox in a different way, the fundamental fantasy is constitutive of (our approach to) reality . . . yet for that very reason, its direct assuming or actualisation cannot fail to give rise to catastrophic consequences. . . . As the common wisdom puts it, a nightmare is a dream come true.[46]

This rupture, between the way things *seem* to appear to me and "the way things 'truly seem to me,' although I never actually experience them in this way"[47] poses a far greater challenge to consciousness than the problem of reconciling our conscious experience with knowledge about the brain. For in the face of a radically reductive scientism that purports to explain such seemingly ineffable experiences as love, happiness, or even consciousness itself, by unveiling their biochemical substrata, I can always stubbornly cling to my subjective, intuitive experience. The standard retort to scientific or philosophical attempts at providing a comprehensive neurobiological account of consciousness is that the "third-person" perspective of science can never account for or get to grips with my "first-person" conscious experience. As a result it is never truly viewed as a serious threat to this experience, because no matter how much knowledge we acquire about the brain, it can in no way be incorporated into our understanding of ourselves or be subjectivized.[48]

We thus find refuge against the threat to our intimate self-experience posed by materialism by holding fast to the truth of what we *feel* as opposed to what we may *know* about the obscure neurochemical processes governing our behavior. This knowledge need have no more of an influence on how we experience ourselves than knowledge about the electromagnetic spectrum affects how we view a Mark Rothko painting. After all, even the most stringent materialist, who refuses the existence of mind and self-consciousness altogether, nevertheless disavows this attitude in her everyday self-relating activity. Although their work may be devoted to exorcising the specter of the "Cartesian theater" (in Daniel Dennett's well-worn phrase) this does not mean that they successfully practice this attitude when they are not theoretically reflecting on it and just going about their daily life.[49] Similarly, many Christians believe that even the most militant atheist does not (could not) *really* believe that the soul will not persist after death, even if he *thinks* he does; in his everyday activity he betrays his unconscious fantasmatic belief that death is not really the end.

What is at stake in the technological revelation of our subperceptual behaviors and unconscious attitudes is thus more radical than the materialist rupture because it specifically *does* bear upon our intimate subjective interiority, sense of self, and the sovereignty of conscious experience. Instead of being simply a question of uncovering the objective chemical processes and physical structures underpinning what we experience while remaining essentially foreign to that experience, it is rather a case of revealing another side to what we think, do, see, and experience which stands in contrast to what we *think* we think, do, see, and experience. This is precisely the upshot of Lacanian psychoanalysis, according to Žižek, in which "I am deprived of even my most intimate 'subjective' experience, the way things 'really seem to me,' that of the fundamental fantasy that constitutes and guarantees the kernel of my being, since I can never consciously experience it and assume it."[50] This, for Žižek, is the status of the Freudian unconscious: the paradoxical notion of an "*inaccessible phenomenon,* not the objective mechanism which regulates my phenomenal experience."[51] This decentering "opens up a new domain of weird 'asubjective phenomena,' of appearances with no subject to whom they can appear: it is only here that the subject is 'no longer a master in his own house'—in the house of his (self-) appearances themselves."[52] The twenty-first century media analyzed by Hansen reveal another, entirely new domain of "appearances with no subject to whom they can appear," as we saw earlier. Such appearances can be "experienced" (by consciousness) *only after they have already been "experienced"*: belatedly via technological registration. In the lived present of the appearances taking place there is nobody "there" to experience them because they occur at temporal scales far below the threshold of conscious perception.

As the technologies of brain scanning and cognitive imaging continue to advance we are faced with the prospect of the real-world arrival of what Žižek calls the *Id-Machine*,[53] a familiar science-fiction device involving an alien intelligence or technology that gives substantial reality to our deepest unacknowledged desires and fantasies. Perhaps the finest example of this in cinema is Tarkovsky's other science-fiction masterpiece, *Solaris*, which takes place almost entirely on board a space station orbiting the planet of the film's title. Due to the strange powers that the planet exercises over the crew members, their inner thoughts, desires, and wishes are materialized as solid entities. However, rather than bringing them joy or happiness, these apparitions are experienced as horrifying and monstrous, sending all of the crew into emotional crises. In the case of the lead character, a psychologist called Kris, his dead wife appears, manifesting his guilt-feelings over her suicide. Another filmic example of the Id-Machine referred to by Žižek is presented in the 1997 Hollywood film *Sphere*, directed by Barry Levinson. It follows a group of scientists after the discovery of an enormous spaceship a thousand feet below the surface of the Pacific Ocean containing a mysterious, perfectly formed metal sphere that can reach into one's mind and manifest all one's worst fears and nightmares. In Žižek's analysis, the eponymous Sphere "is nothing in itself—a pure medium, a perfect mirror that does not mirror/materialise reality but only the real of the subject's fundamental fantasies."[54]

According to Žižek, there is an essential irreconcilability between our true desire and its articulation in a concrete plea or demand, and so that which we truly desire is never actually announced in the explicit wish we were able to express. Consequently "we never truly desire what we wish for or will—for that reason, there is nothing more horrible—more undesirable, precisely—than a Thing that inexorably actualises our true desire."[55] As Writer describes it in *Stalker*, this innermost desire or wish is "not what you would holler at the top of your voice" but what would make you recoil with displeasure if it were presented to you. However, we would fail to recognize ourselves in this distorted reflection:

> Communication with the Thing [that otherness within ourselves that *is* ourselves] thus fails not because it is too alien, the harbinger of an Intellect infinitely surpassing our limited abilities, playing some perverse games with us whose rationale remains forever outside our grasp, but because it brings us too close to what, in ourselves, must remain at a distance if we are to sustain the consistency of our symbolic universe.[56]

We must be careful, however, to avoid any suggestion that such repressed desires, tendencies, instincts, and subperceptual behaviors constitute one's true self: who we *secretly are.* If, to return to the earlier example, one was revealed by brain scan or behavioral sensors to bear hidden racial prejudice, this does not negate the (conscious or unconscious) efforts to self-regulate and counterweigh these tendencies. Rather than exhibiting an enduring "inner self," what such data-mining of our behavioral and cognitive activity reveals, on the contrary, is that the self is nothing but an ongoing, provisional negotiation of competing desires, attitudes, and feelings, of which the conscious part is only one strand among many. My own Catholic upbringing still influences my attitudes, behavior, and expectations despite my conscious efforts to dispel it. It would be wrong, however, to say that as a result of this I am "really" Catholic without knowing it and in spite of my avowed atheism. This projects an utterly false all-or-nothing criteria on tendencies that are characterized by mutable, fluid gradations.

Desires or urges become repressed when there is a conflict of interest with other desires, such that its satisfaction would "cause pleasure in one place and unpleasure in another."[57] What Freud calls *primal repression* is the first phase, "which consists in the psychical (ideational) representative of the instinct being denied entrance into the conscious."[58] This establishes a fixation, and the repressed ideational content "persists unaltered from then onwards and the instinct remains attached to it."[59] The second phase of repression affects all those processes and trains of thought that derive from or have been brought into association with the object of this primal repression. When presented to the patient these processes "are not only bound to seem alien to him, but frighten him by giving him the picture of an extraordinary and dangerous strength of instinct."[60] While it may have been forgotten as far as our conscious self is concerned, it continues to proliferate out of sight of consciousness, entangling more and more thoughts and associations into its web. Of course, the revelation of unsavory desires such as violent sexual urges or hostile or libidinous feelings toward family members are the central currency of psychoanalysis. Nevertheless, those who enter into analysis already feel some psychical conflict, and all the work of the analyst in overcoming resistance prepares the analysand for whatever may be disclosed. The Id-Machine, on the other hand, would meet us unprepared and ill-equipped to come to terms with what it reveals. This would be an automated, instantaneous, and involuntary exposure as opposed to patient preparation.

The difference, Freud suggests, between an unconscious and a conscious presentation is not that they are registered in different areas of the psyche, or different neural regions of the brain. Rather, it is that the conscious content is made up of "the thing plus the presentation of the word belonging to it, while the unconscious presentation is the presentation of the thing alone."[61] So what repression withholds from the prohibited content is its translation into language: the name or "word-presentation" that would correspond to it. Without the hypercathexis provided by the name it will remain in a state of repression. It is word-presentations that bring under control and domesticate unbearable thing-presentations, keeping them at a safe distance. However, contrary to Freud's denial of the neural basis for psychical functions, cognitive neuroscientist Michael Gazzaniga has shown through his studies of split-brain patients (where the left and right hemispheres of the brain have been severed from each other) that this function of interpreting stimuli and forming beliefs about the world is in fact performed by a particular region of the brain: specifically, the left hemisphere, which is responsible for language and speech. What Gazzaniga and his team found is that if information is given only to the right hemisphere through the left ear, for example "raise your arm," the patient will carry it out but will not consciously know why, much like in hypnosis. If the patient's left hemisphere is then asked why he has done this he will invent a belated justification to explain it as much to himself as to the questioner: "I was just . . . stretching my arm." As Gazzaniga explains, "the left hemisphere made up a post-hoc answer that fit the situation."[62] Language thus enables us to make sense of our environment and of our own thoughts and actions, structuring our reality.

Unconscious processes, says Freud, never rise *directly* into consciousness; they are always mediated by the preconscious interposition of word-presentations. The process of analysis works to aid the activity of the preconscious by supplying these intermediate links. The Id-Machine, however, would have the capability to sidestep this work and conjure the unpresentable Thing itself directly, like the apparitions aroused by the Sphere or the sentient ocean on Solaris. Faced with this intrusion, but without either the work of analysis to prepare us for it or the word-presentations which have been denied it, we would lack the tools with which to understand and integrate it into our symbolic framework. As Žižek suggests, it would potentially shatter our fantasmatic consistency: those "auxiliary constructions" we implicitly rely on in order to regulate our reality and at the same time understand and keep in check our experiences and feelings, holding at bay overly powerful stimuli and impulses. Žižek reads the attacking of the eponymous birds in the Hitchcock film as such a tear in the fabric of reality due to overly intense maternal incestuous energy.[63] The birds themselves are a symptom, materializations of the mother's overbearing love for her son, Mitch, and her rage at the disturbance of their family unit by the arrival in the town of Melanie, Mitch's love interest. The symptom is always an unwelcome intrusion, caused by overly strong libidinal cathexis that can find no other outlet since its discharge has been blocked by the preconscious censor. So it is a substitute satisfaction, experienced as suffering, which alerts the ego to the presence of overly charged repressed contents. The Id-Machine would present us with mechanical symptomatic manifestations, but rather than a surrogate satisfaction, which still conceals that which it announces, it would bring to light the repressed material *itself.*

The Timeless Unconscious

We saw in the previous chapter how Kant prohibits the possibility of pure origination from nothing "because its very possibility would already annul the unity of experience." This unity is founded on the consistency of the time series, which gives order and stability to the flux of consciousness. For any radically new event to occur within our experiential space it would have to appear on this temporal horizon and connect with what went before in a relation of succession and so vitiate its quality of absolute novelty. Otherwise, such an impossible occurrence would lead to the disastrous breakdown of the unity necessary for experience. And here the logic is repeated on a psychological level. The repressed contents manifested by the Id-Machine can only appear to consciousness if they are connected with word-presentations and so integrated into our framework of intelligibility. If this fails their arrival would shatter the consistency of our reality leading once again to the breakdown of experience.

This radical incommensurability between the contents of the unconscious and our conscious experience is further amplified by Freud's frequent, puzzling assertions that the unconscious is "timeless." This discovery is presented as an exception to Kant's transcendental conditions of experience. If such is the case the unconscious would apparently represent a site of radical novelty and the direct intrusion of its contents upon consciousness would pose even more of a threat to the consistency of our experience than we envisaged. However, before we could make such a claim we would need to know exactly what Freud means when he says that the unconscious is timeless, and in what sense this amounts to a departure from Kant.

References to the timelessness of the unconscious, which Freud calls "a subject which would merit the most exhaustive treatment,"[64] are scattered across his corpus but without ever being given the exhaustive treatment the subject apparently demands. The earliest explicit reference occurs in 1901, in a footnote to the *Psychopathology of Everyday Life*:

> When traces of memory are repressed they can be shown to have undergone no change over quite a long period. The unconscious is timeless anyway. The outstanding and most surprising character of psychic fixation is that all impressions are retained just as they were absorbed, and moreover are retained in all the forms they assumed in further developments, a relationship which cannot be explained by any comparison from another sphere. According to this theory, therefore, every earlier stage of any material contained in the memory would be available for recollection, even if its elements have all long since changed their original connotations for later ones.[65]

At this early stage it is essentially an expansion upon the thesis of the indestructibility of psychical traces that Freud described one year earlier in *The Interpretation of Dreams*. On this account the unconscious acts like a cryogenic vault, preserving and sustaining impressions, so that "a humiliation that was experienced thirty years ago acts exactly like a fresh one."[66] Rather than the troubling term "timeless," a more appropriate word would be "ageless," meaning the passage of time will neither erode nor alter the impressions.

However, here there is a significant addition: it is not only the *original* impression and the accompanying emotions that are retained but "all the forms they assumed in further developments." So every subsequent recollection of this humiliation, along with every new association or retroactive alteration that it later underwent is preserved alongside it, such that it becomes meaningless to speak of an "original" impression.

Another significant reference to the timelessness of the unconscious occurs in *Beyond the Pleasure Principle*, nineteen years after *The Psychopathology of Everyday Life*, where Freud elucidates the thesis further, apologizing for the obscurity of his remarks, but once again declining to explicate further:

> As a result of certain psychoanalytic discoveries, we are today in a position to embark on a discussion of the Kantian theorem that time and space are "necessary forms of thought." We have learnt that unconscious mental processes are in themselves "timeless." This means in the first place that they are not ordered temporally, that time does not change them in any way and that the idea of time cannot be applied to them. These are negative characteristics which can only be clearly understood if a comparison is made with *conscious* mental processes. On the other hand, our abstract idea of time seems to be wholly derived from the method of working of the system Pcpt.-Cs. And to correspond to a perception on its own part of that method of working. This mode of functioning may perhaps constitute another way of providing a shield against stimuli. I know that these remarks must sound very obscure, but I must limit myself to these hints.[67]

As is fleetingly referred to here and expanded upon slightly in the "Note Upon the 'Mystic Writing-Pad,'" experiential time is a product of the perceptual system. It is a result of the periodic sending forth and withdrawal of cathectic innervations, likened by Freud to a sending out of "feelers" toward the external world which withdraw as soon as they have felt or tasted the excitations that are there. As soon as this cathexis withdraws, consciousness withdraws along with it. The periodic fluctuations in the current of innervations, akin to a continuous writing and raising of the sheet on the famous Wunderblock, is, according to Freud, the origin of our concept of time and temporal succession. However, it is still unclear at this point what Freud intends by the assertion that time itself acts as a shield against stimuli. In *The Ego and the Id* we are presented with another part of the picture. The process of temporalization is said to coincide with reality testing, so bestowing a temporal order upon an internal perception belongs to the same perceptual function as the assessment as to whether or not it is real.[68] Thus at the same time as we ascertain whether an internal psychical image corresponds with objective reality we also assign it a place in time: if it no longer confronts us from the outside, it therefore belongs to the past and we can begin to distance ourselves from it.

Freud addresses this process by which we consign impressions to the past in "The Dissection of the Psychical Personality" from the *New Introductory Lectures on Psychoanalysis*, where he gives perhaps his most extensive treatment of the theme of the timeless unconscious, but once again accompanied by the regret that "too little theoretical use" has been made of it:

> [We] perceive with surprise an exception to the philosophical theorem that space and time are necessary forms of our mental acts. There is nothing in the id that corresponds to the idea of time; there is no recognition of the passage of time, and—a thing that is most remarkable and awaits consideration in philosophical thought—no alteration in its mental processes is produced by the passage of time. Wishful impulses which have never passed beyond the id, but impressions too, which have been sunk into the id by repression, are virtually immortal; after the passage of decades they behave as though they have just occurred. They can only be recognised as belonging to the past, can only lose their importance and be deprived of their cathexis of energy, when they have been made conscious by the work of analysis. . . . Again and again I have had the impression that we have made too little theoretical use of this fact.[69]

So it is only by determining an event as belonging to the past, whether this be a sense impression or a wishful fantasy that has been registered but never acted upon, that it can be defused and its impact softened. Without this temporalization the event will continue to insist itself upon us in the present. This corresponds with recent experimental clinical treatments for sufferers of post-traumatic stress disorder (PTSD). PTSD occurs when the psychical imprint of a distressful event was so strong that the memory refuses to fade with the passage of time, remaining as potent as on the day it occurred. The treatment consists in the patient being administered beta blockers shortly after the traumatic memory has been reactivated. Scientists believe that the drug blocks the reconsolidation of the traumatic memory by inhibiting proteins in the connections between brain cells, which in turn induces a toning down of the emotional coloring of the memory so that it is not forgotten altogether but allowed to become less intense.[70] In Freud's terms this deprives the memory of its cathexis of energy, allowing the patient to consign the event to the past rather than living it in the present, but the result is chemically induced rather than via the talking cure of analysis.

Both of the last two cited passages began with a reference to Kant and go on to outline the psychoanalytic discovery of an exception to the a priori conditions of experience, but is this really an exception to Kant's system? Freud misrepresents Kant in stating that time is a necessary form, first of "thought" and then of our "mental acts." Rather, time, for Kant, is the pure form of inner intuition and the condition of all phenomenal appearance. For any event to occur, whether it be outside of us in phenomenal space or an internal process, it must take place upon this horizon. Furthermore we have just seen how Freud himself insists that as soon as the unconscious traces are made conscious they are assigned a place in time, for in order for us actually to perceive these processes they must be placed in a personal chronology. So for Freud, too, perception necessarily takes place under temporal conditions. Freud's addition to this picture is his contention that psychical processes that are not being perceived are still active in the mind, and that time has no jurisdiction over them. We can have no possible notion of what form such contents may take while existing in this lawless, atemporal state, for any possible attempt at description must rely on the notion of simultaneity, which itself is of course a temporal modality.[71]

Where Freud's psychological account differs further from Kant's transcendental account is in the assertion that the activity of the perceptual apparatus is responsible not for time as such (transcendental ideality), but for our *abstract idea* of time. This is no doubt equivalent to what Heidegger calls the vulgar concept of time: the calculation of divisible units of time "in which" events take place according to the linear succession of past—present—future, and centered around the primacy of the now. In this sense, perhaps, as Derrida proposes, we should read Freud the way Heidegger read Kant, and say that "the unconscious is no doubt timeless only from a certain vulgar conception of time."[72] On such a reading the unconscious would not be withdrawn from temporality as such, but rather fall outside of what Heidegger calls "within-time-ness." Derrida suggests that the question concerning the timelessness of the unconscious should be posed in relation to the malleability of mnemic traces, exemplified by the potential for a previously innocuous event to take on traumatic proportions some years later, retroactively constituting itself as the cause of a later complex (the phenomenon known as *Nachträglichkeit,* of which the Wolf-Man is the key case study). Because the past that we experience in memory is a reconstruction, of a present that was itself already reconstituted and never "present," it is not hard-coded into our psychic apparatus but forever being produced, altered, and rewritten. As such the time sequence is endlessly fluid and flexible. It would therefore be a meaningless question based on metaphysical presuppositions to ask about the form taken by such timeless traces when they are outside consciousness.

So Freud's thesis of the timelessness of the unconscious should not be seen as an exception to Kant's transcendental conditions of experience, and certainly not the site of an overcoming of the finite limits of subjectivity. Rather, if we follow Derrida, this entangling of different temporal states, where every act of recollection transforms that which it recollects, is finitude itself. Furthermore Freud often insists, recalling Kant's account of the productive imagination, that all the materials that make up the "contents" of the unconscious are derived from experience, and so, in the eyes of a Kantian, can be traced back to a sense impression which was subject to the transcendental conditions of appearance. This reliance on receptivity is precisely the impasse of finitude that we encountered previously.

However, in his later work Freud seems to violate this rule as he comes to hold the view that traces remain in our psyche which cannot have been derived from personal memory but are rather archaic psychical vestiges which have expressed themselves in fantasy. As Freud explains,

> I believe these *primal fantasies*, as I should like to call them . . . are a phylogenetic endowment. In them the individual reaches beyond his own experience into primeval experience at points where his own experience has been too rudimentary. It seems to me quite possible that all the things that are told to us today in analysis as phantasy—the seduction of children, the inflaming of sexual excitement by observing parental intercourse, the threat of castration (or rather castration itself)—were once real occurrences in the primeval times of the human family, and that children in their phantasies are simply filling in the gaps in individual truth with prehistoric truth.[73]

This notion of psychic inheritance complicates matters significantly, suggesting that in addition to "timeless" traces of individual memories the unconscious also consists of fantasmatic enactments of primordial experiences that were never lived by consciousness. Our psychical range therefore stretches back far beyond the period of one's own individual chronology. This would be a prepersonal experience that must by nature be distributed across psyches within a given cultural milieu since we each share a history and ancestry.[74] These "primal fantasies" that Freud is compelled to assume, as he himself admits,[75] bear a striking, somewhat surprising, resemblance to the theory of the archetypes of the collective unconscious as expounded by Freud's most famous disciple turned apostate, Carl Gustav Jung. However, Jung goes further than Freud in his break with Kant and in doing so troubles the firm separation between internal and external that Freud's entire model of the psyche rests on.

Jung and the Collective Unconscious

Jung distinguishes between two orders of the unconscious psyche: the personal unconscious, consisting of forgotten or repressed memories, traumas, fantasies, and impulses, which is the object of Freudian psychoanalysis, and the *trans*personal—collective—unconscious, which is "detached from anything personal and is entirely universal" and whose "contents can be found everywhere."[76] The assumption of an inherent tendency toward the production of universal psychic images is a necessary one, says Jung, if we are to be able to account for the recurrence of myths, legends, and religious symbolism in kindred forms all over the world and throughout history. When prepersonal mythological images and symbols appear in fantasies and dreams we are no longer dealing with the individual's own memories or experiences, as in the phenomenon of cryptomnesia, but with "the deeper layer of the unconscious where the primordial [archetypal] images common to humanity lie sleeping."[77]

Jung sees the unconscious as performing a compensatory function, developing as counterbalance to the ego out of neglected aspects of one's own personality. If, for example, we are excessively ordered and clear-headed in our daily lives, the irrational, passionate side of ourselves that we neglect will not be quashed but will proliferate, attaining more and more autonomy the more it is neglected. This leads to an imbalanced psyche that will eventually reach an impasse. It is at such moments of impasse, Jung says, that archetypal images most commonly force themselves upon the subject, offering a potential release that was not previously available. An encounter with the collective unconscious is, for Jung, responsible for every variety of spiritual conversion or vision, as well as abrupt midlife changes of career or lifestyle. The archetypes are specially charged with psychic energy and their emergence in consciousness brings with it an overwhelming affective resonance, inhibiting all other mental contents. As Jung describes it, "[initially,] this always feels like the end of the world. . . . One feels delivered up, disoriented, like a rudderless ship that is abandoned to the moods of the elements. So at least it seems. In reality, however, one has fallen back upon the collective unconscious, which now takes over the leadership."[78] Of course, not every such encounter with the collective unconscious leads to this rewarding recalibration of one's life. In some cases it leads to a ruinous collapse from which there is no way back to one's prior stability.

Nevertheless, in spite of its dangers it is only after a reckoning with the collective unconscious that we approach wholeness or "come to selfhood," for the "self" in Jung's terminology is not a given but the outcome of a process of individuation. What we are as a *self* transcends that which can be brought to conscious introspection. Or, as Jung writes, the "part" is constitutively unable to comprehend the whole.[79] In itself and in principle, the self is a holistic unity lacking nothing, but we will never consciously attain the knowledge of this whole, since it infinitely exceeds our imaginative capabilities: "however much we may make conscious there will always exist an indeterminate and indeterminable amount of unconscious material which belongs to the totality of the self."[80]

This, then, is the nub of the subtle but crucial difference between the Freudian and Jungian models of the psyche. The Freudian unconscious can never be exhausted because it is that which forever escapes our attempts at self-mastery, dividing one from oneself and precluding immediate presence to self.[81] For Jung, on the other hand, the unconscious will never be exhausted because it carries within it unfathomable depths which are essentially beyond our powers of comprehension. Both place an emphasis on our finitude in the face of radical alterity but while for Freud the subject is barred from wholeness because of an irreducible subtraction or fissure, for Jung we are denied wholeness because of an inexhaustible excess.

It should be clear from all of this that Jung would present a very different answer to the question of the "oceanic" to that given by Freud. Instead of being a remnant and intimation of the all-encompassing ego of early childhood, this feeling of limitlessness would instead be the result of an encounter with the collective unconscious and an experience of the exponential surplus of the *self* over the ego. No matter how introspective we may be and how well we may know ourselves we are all capable of wider consciousness, since "it is highly probable that we are still a long way from the summit of absolute consciousness."[82] The Hegelian flavor of this claim is obvious, but the sentiment is actually rather more Kantian, for Jung elsewhere says that "no mortal mind can plumb the depths" of the collective unconscious. The most we can do is proceed according to "the hypothetical 'as if,' "[83] which is an unmistakably Kantian formulation: an infinite striving toward a regulative ideal we know to be unattainable, nevertheless proceeding *as if* the sequence was complete in itself.

The inevitable question is whether increased understanding of the brain may put us in a better position to "plumb the depths" of the collective unconscious and bring us closer to the absolute consciousness that Jung says we are structurally prohibited from attaining. However, as Jung repeatedly insists in the face of a recurring criticism, the archetypes of the collective unconscious are not innate ideas residing within us and waiting to be uncovered by whatever means. Rather, they are forms of possible experience which carry no actual content prior to their "conscious elaboration."[84] In spite of a frustrating terminological inconsistency throughout his corpus which encourages confusion, Jung draws a clear and unequivocal—indeed Kantian—distinction between the archetypes themselves and archetypal ideas, which are their conscious derivatives.[85] The former are hypothetical and irrepresentable and their existence can only be inferred from their effects, which is the way they are expressed in the life of the individual. This expression (in dreams, or fantasies, and so forth) is inevitably shaped by the subject's personal history and circumstances although it generally proceeds according to familiar patterns. So rather than images lying deep within the unconscious, they are potential patterns of behavior making their appearance "only in the course of amplification."[86] The therapeutic process Jung calls "active imagination" is the forcing or helping on its way of this course of amplification, where the subject delves into the unconscious psyche by way of active, spontaneous fantasy. On following these fantasmatic inventions where they lead of their own volition, they invariably follow certain grooves or psychic imprints, where familiar mythological figures and motifs appear that betray their archetypal character. But it is important to stress that prior to this amplification process there is nothing there to access.

So it would make no more sense to think we could uncover the archetypes of the collective unconscious in their raw form by examining the brain than it would to suggest we could access the pure categories of the understanding by such means. Rather than innate or inherited experiences the archetypes are formal structures guiding possible experience. However, unlike Kant's a priori conditions of possible experience, which are necessary and unchanging, the archetypes are dynamic and shaped by past experience. As Jung describes, they are made up of "not only the residues of archaic, specifically human modes of functioning, but also [of] the residues of functions from man's animal ancestry, whose duration in time was infinitely greater than the relatively brief epoch of specifically human existence."[87] For this reason no exhaustive taxonomy can be drawn up of the archetypes such as Kant does with the categories. They are a posteriori deposits of ancestral experience that have passed over into quasi–a priori conditions of future experience. In Jung's words, the collective unconscious is at once "the repository of man's experience" and "the *prior condition of this experience*,"[88] thus disturbing the relation of cause and effect in a similar way to the Freudian *Nachträglichkeit*. As mentioned above, the constellating of archetypal ideas in the unconscious takes place as compensation for neglected aspects of the subject's psyche, so that when they build up an irresistible force they impose themselves on the life of the subject, forcing them in a certain direction so as to reorient their psychic balance. "Perhaps—who knows?—these eternal images are what we mean by fate."[89] However, despite their formal, a priori nature, given their empirical genesis the question as to where exactly they are registered becomes more pressing than it does for the Kantian a priori. If the archetypes are not to be found in the brain then where are they located?

Synchronicity and the *Unus Mundus*

Jung equates the irrepresentable nature of the archetypes with the smallest particles that physics deals with, whose nature can only be known by their effects. In both cases the physicist or psychologist is attempting to define an order of nature whose behavior is altered by the fact of its being observed and can at best build up a probable model of how these quantities behave based on their observable effects. Jung further hypothesizes that, given the presence of two classes of entity or property whose existence must be assumed but which cannot be represented or shown in person, "there is always the possibility—which we tend to overlook—that it may not be a question of two or more factors but of one only":

> The identity or non-identity of two irrepresentable quantities is something that cannot be proved. . . . Since psyche and matter are contained in one and the

> same world, and moreover are in continuous contact with one another and ultimately rest on irrepresentable, transcendental factors, it is not only possible but fairly probable, even, that psyche and matter are two different aspects of one and the same thing. (214–15)

Jung uses the term *psychoid* (always as an adjective, never a substantive) to describe these irrepresentable psychophysical processes. This "one and the same thing" which is neither psychic nor material in nature but antecedent to their differentiation, is named, using the terminology of alchemy and medieval philosophy, *unus mundus*, meaning "one unitary world." Jung felt this hypothesis to be far from the obscure mysticism it can appear to be at first sight but to be a legitimate response to empirical data, informed by developments in particle physics. Indeed Jung developed this concept in collaboration with the famous physicist and Nobel laureate Wolfgang Pauli (a relationship generally passed over in silence in physics circles, or excused as the individual eccentricities of a great man that has no bearing on his work.)

For Jung there are certain privileged events or experiences which index this irrepresentable unity of psyche and world and which carry profound and far-reaching implications. Such events are those statistical anomalies, commonly attributed to chance or coincidence, which seem to fall outside of any known causal relation and so elude rational explanation. A classic example is the case of the Swedish spiritualist Emanuel Swedenborg's dramatic vision of the great fire of Stockholm in 1759 while he was dining in Gothenburg, two hundred fifty miles away. According to the eye-witness accounts it was only two days later that reports from Stockholm confirming Swedenborg's vision, down to the smallest detail, reached them in Gothenburg. Another famous example is one of Jung's own, from his analytic experience. A young patient of his was describing a dream in which she was presented with a golden scarab. In the middle of her account Jung became aware of a tapping noise against the window of his practice and opened the window, through which flew a rose-chafer beetle, "the nearest analogy to a golden scarab that one finds in our latitudes . . . , which contrary to its usual habits had evidently felt an urge to get into a dark room at this particular moment."[90]

Although the two cases are very different in nature—for one thing the patient was not presenting her dream as if it were a prophetic vision—both are examples of the phenomenon Jung calls *synchronicity*, defined as the "coincidence of a psychic state with a corresponding objective process" (480). More everyday, commonly experienced instances are unlikely meaningful coincidences, such as thinking of a friend from whom one has not heard in a long time immediately before receiving a telephone call from that same friend, or successions of chance events such as a number or word recurring again and again throughout the course of a day or number of days (a phenomenon Freud addresses in *The Uncanny*.) Several further examples, clinical and nonclinical, involving demonstration of seemingly impossible knowledge about parallel or future events are adduced by Jung in support of the synchronicity thesis.

Clearly, from the perspective of the positive sciences there is nothing here to explain because we are dealing either with chance phenomena, of admittedly low probability, superstitiously assumed to carry significance, or inadvertent falsification of memory (or outright fraud in the case of Swedenborg.) Simply put, instances such as these do not constitute an object for scientific knowledge. However, rather than explaining them by dissolving them Jung is concerned to preserve their psychical significance and to understand how separate events (psychic and physical) can be related to each other meaningfully without that relation being a causal one. Just as Freud found meaning in apparently meaningless dreams and parapraxes, so Jung reads meaning into seemingly meaningless coincidental connections of events.

Of course, since these are singular, anomalous occurrences they are on principle incapable of being premeditated and examined in controlled conditions, for the experimental method by nature aims at establishing regular, repeatable events and thus ruling out of consideration the unique or rare results that are put down to chance deviations. Causality, says Jung, is a *statistical* truth, not an absolute truth, and is only *generally* valid, when operating on the macrocosmic scale. "In the realm of very small quantities *prediction* becomes uncertain, if not impossible, because very small quantities no longer behave in accordance with the known natural laws" (421). So while in general the course of nature can be unfailingly expected to follow a causal course, when we are dealing with particular events on a micro scale we can never predict the outcome with absolute certainty. Voicing a commonly raised objection to the limitations of the scientific procedure, Jung contends that the answers obtained in experimental conditions are largely shaped by the questions posed, thus giving only a partial, statistical, or average view of the natural world. However, Jung radicalizes this relatively uncontroversial claim, drawing the conclusion that since causality is not a universal truth there must be connections of events which are *acausal*, which thus demand another "connecting principle" to account for them. With synchronicity we are not dealing with a relation of cause and effect, but rather a "falling together in time, a kind of simultaneity. Because of this factor of simultaneity, I have picked on the term 'synchronicity' to designate a hypothetical factor equal in rank to causality as a principle of explanation" (421). It is important to note the crucial term *hypothetical* here, and the concomitant sense that synchronicity for Jung does not constitute a positive addition to knowledge but rather acts as a regulative model to help guide an interpretation of seemingly unaccountable phenomena.

Like Freud, Jung refers to Kant's transcendental conditions of sensibility and suggests, albeit in a very different way to Freud, that these conditions do not govern unconscious processes:

> They [time and space] are hypostatised concepts born of the discriminating activity of the conscious mind, and they form the indispensable coordinates for describing the behaviour of bodies in motion. They are, therefore, essentially psychic in origin, which is probably the reason that impelled Kant to regard them as *a priori* categories. But if space and time are only apparently properties of bodies in motion and are created by the intellectual needs of the observer, then their relativisation by psychic conditions is no longer a matter for astonishment but is brought within the bounds of possibility. This possibility presents itself when the psyche observes, not external bodies, but *itself.* (436)

In the case of Swedenborg's vision, it was not a case of the mystic "seeing" the actual fire raging in Stockholm through some temporarily extended faculty of perception, for the information did not reach him from the *outside* but from the *inside.* Thus for Jung the distinction between internal and external is not so firmly established as it was for Freud, and as a result, "reality testing" becomes a far less straightforward matter. These inner processes can become consciously registered in situations of extreme agitation and emotional intensification where archetypal contents are constellated, and it is invariably with the archetypes that we are dealing in synchronistic phenomena. The dream of the golden scarab, for instance, occurred at a critical moment of deadlock in the patient's treatment, and the scarab, Jung tells us, is a familiar archetypal symbol of rebirth. As the collective unconscious is universal and everywhere the same there is the ever-present possibility that what is taking place at any one time in the collective psyche of an individual is "also happening in other individuals or organisms or things or situations" (481). This is what seems to have occurred in the scarab dream; it was a conscious representation deriving from the causally inexplicable unconscious knowledge of the events of the following day's session with her doctor.

What experiences such as these bear witness to, according to Jung, is a form of "'knowledge,' or 'immediacy' of psychic images" (446) which does not derive from sense perception. The conscious interpretation of this unconscious knowledge arrives within the subject like any other spontaneous thought and can only be verified as a synchronistic occurrence after the physical event has been perceived. This suggests that there may be many such examples of this inexplicable knowledge which are never recognized as such because the physical event with which it corresponds is not witnessed by the person to whom it has appeared. Many of us must have experienced at some time or another that uneasy sensation of ominous precognition, and there are numerous stories, an example of which is given by Jung, where a person claims to have "sensed" or known when a loved one has died. What happens in such cases "is a kind of *creatio ex nihilo*, an act of creation that is not causally explicable" (480). This is something we were led to rule out as inconceivable in the Kantian terms of our previous chapter. The two impediments to such a notion were, first, that every psychical image or impression derives from sensibly given material and second, that the necessary consistency of the time-series precludes the emergence of something that cannot be traced back to a preceding cause. Both objections are sidestepped in the light of synchronistic phenomena, the first by the inexplicable nonsensible knowledge such experiences exhibit and the second by the "psychic relativity" of time and its abolition in the unconscious.

So what Jung is "finally compelled to assume" is that "there is in the unconscious something like an a priori knowledge or immediate presence of events which lacks any causal basis" (447). If such events were a case of causality then either the dream or vision which foresees a future or simultaneous event, caused the event to take place in some telekinetic way, or the physical event caused the psychical process, retroactively positing itself somehow. "In either case," says Jung, "we come up against the unanswerable question of transmission" (483). This question of transmission is none other than that of the two possible relationships explaining the correspondence between object and representation that Kant gives in the letter to Marcus Herz. For Kant, as we know, in finite intuition the object is the cause of the representation while in the divine intuition it is the other way around. What Jung does, however, is to displace the terms of the question by presenting a *third* alternative that Kant did not, and indeed could not, have considered.

This third alternative relies on the *unus mundus* hypothesis, suggesting that the two related terms—the psychical experience and the physical event—take place on *another plane prior to their differentiation*, and both the knowledge and the event itself are its respective manifestations. In other words, "the same living reality [is] expressing itself in the psychic state as in the physical" (452). So with regard to the two alternatives Kant poses, this would be neither a receptive nor a productive intuition, but still nevertheless a form of intelligible correspondence between thought and being that is not mediated through the senses. As such the problem of transmission is circumvented, but what exactly forms the correspondence between the two states if it is not a case of causation? Jung's answer to this question postulates an a priori "meaning" or equivalence, which exists independently of the psyche:

> If—and it seems plausible—the meaningful coincidence or "cross-connection" of events cannot be explained causally, then the connecting principle must lie in the *equal significance* of parallel events; in other words, their *tertium comparationis* is *meaning*. We are so accustomed to regard meaning as a psychic process or content that it never enters our heads to suppose that it could also exist outside the psyche. But we do know at least enough about the psyche not to attribute to it any magical power, and still less can we attribute any magical power to the conscious mind. If, therefore, we entertain the hypothesis that one and the same (transcendental) meaning might manifest itself simultaneously in the human psyche and in the arrangement of an external and independent event, we at once come into conflict with the conventional scientific and epistemological views. (482)

For Jung this is, in fact, the least mystical explanation that does justice to the empirical data without attributing to the psyche "a power that far exceeds its empirical range of action," namely intellectual intuition. However, by avoiding attributing *this* particular supernatural power to the psyche, Jung risks ascribing to it another, equally exorbitant faculty. For when the threshold of consciousness is sufficiently lowered so that unconscious, archetypal contents can penetrate into our conscious mind this can grant us access to what Jung calls, in quotation marks for caution, " 'absolute knowledge,' " pointing to "the presence in the microcosm of macrocosmic events" (489). The microcosm here, which like the Leibnizian monad reflects the whole of reality, is the collective unconscious.

This is speculative language that must inevitably appear somewhat fantastic because it aims to render intelligible to consciousness something which is necessarily inconceivable to it. So in the case of the scarab dream, for instance, we are not dealing with paranormal foreknowledge, or still less psychokinesis, but with two distinct manifestations of the *same event* that are connected by "meaning" or significance.[91] Since in the unconscious psyche time and space no longer apply and "knowledge finds itself in a space-time continuum in which space is no longer space, nor time time," then if the unconscious should "develop or maintain a potential in the direction of consciousness, it is then possible for parallel events to be perceived or 'known' " (481). This knowledge cannot be employed at will since synchronicity events are by their nature rare and incapable of being premeditated, but as scientific knowledge about the workings of cognition advances perhaps it could eventually be in a position to definitively test Jung's hypothesis and discover such a parallelism between irrepresentable psychic and physical processes. More to the point, if the collective unconscious is indeed a resource of "absolute knowledge" perhaps this could conceivably be harnessed and exploited technologically, overcoming our previous reservations and, in the words of Žižek, render a faculty of infinite cognition (or absolute knowledge) "potentially available to all of us."

However, just as it was with the archetypes, such a notion is precluded in principle, for the kind of knowledge displayed in synchronistic events "has nothing to do with brain activity" (505). For Jung, the psyche is not exclusively localized to cognitive activity, it rests also on "a nervous substrate like the sympathetic system, which is absolutely different from the cerebrospinal system in point of origin and function, [and which] can evidently produce thoughts and perceptions just as easily as the latter" (510–11). Jung illustrates this with anecdotal accounts of comatose patients displaying awareness of events going on around them and giving detailed reports of what they saw upon regaining consciousness, as well as with the behavior of bees, for instance, which in their much-discussed communicative movements (or dances) display "transcerebral thought and perception" (511).

Nonetheless, even if this rich psychical reservoir cannot be directly accessed technologically, Jung does furnish us with a model for rethinking the gap between internal and external which has set the terms of our enquiry so far. Just as in Jung's analytic treatment, where the archetypes are constellated at a point of impasse to disclose a hitherto unthinkable means of escape, so Jung himself shows us a line of advance out of the impasse of finitude. This noncerebral, bodily form of knowledge is, for Jung, an exemplar of the psychoid property inherent in matter so that "thought," broadly conceived, is not confined to the human mind but pervades that which is its "object." Yet this is not a *simple* panpsychism, suggesting that water, plants, or rocks possess a rudimentary form of conscious perception, although it is undoubtedly redolent of it. For such a notion could still be considered a dualism, extending the capacity for thought to inanimate objects while still upholding its exceptional status.[92] Rather, Jung's ontology levels down the disjunction between "mere thought" and positive being, since it is only for a model which strictly upholds such a distinction that the synchronicity phenomena remain inconceivable. These latter do not form a miraculous bridge establishing a momentary sympathetic connection between two isolated properties, but point to a way of reconceiving the relationship itself, appearing to show "that there is some possibility of getting rid of the incommensurability between the observed and the observer." "The result," says Jung, if such is the case, "would be a unity of being which would have to be expressed in terms of a new conceptual language" (512). This, borrowing an expression from Wolfgang Pauli, Jung terms a "*neutral* language": neutral because it does not discriminate between what is inner and what is outer, the knower and the known.

Jung's *unus mundus* concept simultaneously accomplishes two seemingly incompatible things: first, its ontological neutrality abolishes the special status of thought, bringing it down from its lonely tower and is thus consistent with the approach of cognitive science which treats consciousness as a physical phenomenon no different from any other. Second, in doing so, thought is not *reduced* but greatly enhanced and set free from finite limits and its enslavement to receptivity. It may not be intellectual intuition, but, as Meillassoux discovered, one cannot escape from the Kantian system using the tools found within it. Nevertheless, it is necessary to remain consistent with Kant in order to avoid charges of indulging in groundless metaphysical speculation. This, I would argue, Jung does, despite his reputation for mysticism, reimagining the subject-object relation along the lines of a Parmenidean identity of thought and being but without resorting to an idealist privileging of the former over the latter.

An obvious objection presents itself here in the light of our earlier discussion of Quentin Meillassoux: is *unus mundus* not just an elaborate variation on the procedure Meillassoux calls "absolutising the correlation," or Hegelianism?[93] This is Paul Bishop's claim in his book *Synchronicity and Intellectual Intuition in Kant, Swedenborg and Jung,* where he argues that Jung succumbs to a mystic metaphysics in which mind and world form an absolute unity.[94] However, a more careful, sympathetic reading of Jung than Bishop seems willing to give reveals a rather more nuanced thinking to be at work. Far from an absolutized reciprocity of thought and being, this psychoid absolute, as I have tried to insist, takes place *prior to the correlation* and names a stratum of being antecedent to the differentiation into subject and object, thought and the given. Neither side of the relationship can be conceived in such terms since thought is not yet thought and being is not yet *given* to thought. This is why a new "neutral language" is required.

So to return to those two founding moments of Meillassoux's procedure that we discussed in our previous chapter: (1) the immanent point of departure (from *within* the correlation) and (2) the subsequent transgression through logical reasoning (intellectual/dianoetic intuition), it can be seen that Jung adopts a similar procedure. Jung, too, recognizes the necessity of remaining consistent with Kantian limits, while breaking free of them to enable thought to exceed itself and access an absolute independent of and prior to thought. Or in Meillassoux's terms, Jung escapes the correlationist circle from *within* rather than merely positing an autonomous real. Second, just as Meillassoux proceeds to access a primary absolute via the logical explication of an intuition (of facticity), so Jung's absolute is approached via rational demonstration following the intuition of causally inexplicable phenomena. This absolute, however, avoids the problem we found in Meillassoux's system, namely squaring the circle of being at one and at the same time irreducible to, yet exhaustively available to, thought. The infinite excessiveness of being over thought is not reduced, but nor is their irreconcilability affirmed because thought is already part of that which it thinks. So rather than attempting to erect a bridge to connect thought and being while forever keeping them separated by an irreducible chasm, Jung instead offers us a way of escaping the problem. For a bridge would merely be a means of passage or communication between two isolated territories, and this correspondence can only be conceived as a causal relation (as Kant's letter to Herz asserts).

We saw that by following Žižek's lead and asking whether the finite gap between internal and external could be transcended by means of technology we were always going to be thwarted in this effort due to its underlying assumptions. Approaching this relationship *as a gap* forbids in advance any attempt to overcome it. So after having reached a certain impasse we were compelled to attempt to think the internal-external relationship otherwise and so transcend it by alternative means. This was necessary because it became apparent that unless there is a faculty of the mind that transcends the receptivity constitutive of finitude then no neurotechnology can truly carry such radical consequences as they may seem at first sight to present. The rub, of course, is that once we have traversed this abyssal space by an alternative means in order precisely to secure the technology's metaphysical potential we then find that we no longer need the instrument, for that which we had hoped to achieve was, by thinking it otherwise, already achieved. It seems that either way the technical system fails to bridge the gap: if there is this insurmountable finite gap between thought and being then no instrument, however advanced, could fully overcome it. If, on the other hand, we can conceive the relationship rather along the lines of a continuum or neutral space then in effect we already have what we are looking for and the technology is no longer required to perform this function.

On the back of this Jungian argument we could now propose a different conceptualization of the creative process of the artist, in which it is conceived according to the synchronicity model; the relationship between the motivating idea or thought and the resulting object would then need to be reconsidered. Perhaps those rare moments of inspiration are not the subjective, albeit unconscious, creations of the individual mind but examples of acausal psychophysical correspondence. When Mozart speaks of the way musical ideas flow unaccountably into his mind—"[whence] and how they come, I know not; nor can I force them"[95]—might we not say that he experiences the inexplicable "knowledge" or "immediate presence" of that piece of work as a transpersonal process in much the same way as Jung's patient was "visited" by the scarab beetle in a dream? To be an artist then, to create, would thus mean being in possession of a form of ecstatic knowledge; it would mean having access to something (some inspiration or idea) that has positive existence outside of oneself. Much the same could be said for those moments of absolute clarity experienced by the scientist, mathematician, or philosopher, when one feels at last to have *understood* something in a flash of insight. So, recalling the reservations we had about Croce's philosophy in our first chapter, where we noted that Croce only overcomes the gap between idea and event at the cost of giving one term absolute priority over the other, here, by contrast, the gap is overcome without giving in to idealism.

However, such a proposal undoubtedly invites accusations of tying truth to the irrational, and of suggesting that all access to truth is necessarily on the order of revelation or enthusiasm. To this charge I would argue that any unprecedented scientific breakthrough, philosophical insight, or artistic inspiration must depart from simple rule-following, logical inference, or calculation. There must be, as Adorno has said, an essentially unjustifiable *gap* that even the thinker or artist herself cannot fully account for. This experience of understanding, in which a problem or a solution becomes clear, is never simply the inevitable outcome of an implemented program or formula, otherwise it would in principle be known already and merely awaited its eventual computation. So this experience of insight is more of a *soliciting* than a *creating* strictly speaking, similar to when Lyotard writes that "thinking, like writing or painting, is almost no more than letting a givable come towards you";[96] or, when Nietzsche describes the process of inspiration thus: "one hears, one does not seek; one takes, one does not ask who gives; a thought flashes up like lightning, with necessity, unfalteringly formed—I have never had any choice."[97] After all, Kant describes the creative genius as one who does not know from whence his inspiration derives and who cannot produce it at will since it is irreducible to the mere following of rules. Thus at the same time as Kant exalts the artist of genius he simultaneously humbles him by placing him in thrall to something that he no more exploits or understands than the nonartist does. So this momentary insight *is* essentially irrational and undermines the attribution of responsibility, for how can one be truly and wholly *responsible* for an idea that arises involuntarily?[98]

It is this experience of insight or understanding that I am suggesting could be usefully conceived according to the model of synchronicity. However, while in Adorno this gap or leap that constitutes insight is held to be a sign of the irreducible separation of thought from being (as we saw in the Introduction), here we are using it as evidence of their coincidence. Synchronicity allows us to think the act of creativity not as the production of a sovereign agent, but as a transpersonal event or occurrence, thereby giving a form of positivity and independence to the *object* of creation (that which is invented or conceived.) However, this does not claim, or indeed seek, to offer a definitive, reductive explanation of creativity since, as already noted, Jung proposes the idea of synchronicity as a hypothetical, heuristic concept rather than as a positive, demonstrable fact.

However, there remains a further gap to be accounted for that cannot be dissociated from that which separates interiority from exteriority, or thought from being: namely, that between my interiority and another's. For if the apparent irreducibility of the gap between thought and what is given to thought can be overcome by reconceiving this relationship along the lines of a Jungian neutral unity, how are we to accordingly conceive the intersubjective relation between the multiple "sites" or "instances" of thought? Surely the flattening out that Jung's psychoid ontology describes must work horizontally as well as vertically. So does it follow from the *unus mundus* hypothesis that the gap separating *I* from *other* is likewise overcome at this prior level which we described above as being antecedent to the differentiation between subject and object? Or, to put it differently, if when Jung suggests that synchronicity events offer us the "possibility of getting rid of the incommensurability between the observed and the observer," what happens when the observed is another observer? This incommensurability is of an altogether different order and so must be accounted for separately.

The reason this becomes a particularly compelling question in the broader context of this book is that neural interface technologies are already being employed to develop models of brain-to-brain communication where, instead of transferring control information from the brain to an external device, information is relayed directly from one brain to another. This is a new and burgeoning area of research whose implications pose profound and radically novel questions of the intersubjective relationship, fundamentally recasting the old philosophical problem of "other minds." The authors of a 2014 research article describe this shift in dramatic terms: "until recently, the exchange of communication between minds or brains of different individuals has been *supported and constrained* by the sensorial and motor arsenals of our body. However, there is now the possibility of a new era in which brains will dialogue in a more direct way" [my emphasis]. Such technologies, they say, "will eventually have a profound impact on the social structure of our civilisation and raise important ethical issues."[99] The final chapter will be concerned with precisely these questions.

As with our previous enquiries, the problem must be posed in a particular way. For if technology can indeed redefine the relationship between one mind and another and transcend the distance which separates them, what does this possibility tell us about how that relationship is constituted? If, as David Roden speculatively proposed, "the solitude [of mental states] that [is considered] definitive of the human condition can be overcome by Brain-Machine-Brain interfaces that duplicate the functions of our home-grown commissures,"[100] such a possibility demands nothing less than a radical reassessment of the very nature of intersubjectivity along with a thoroughgoing analysis of the implications such a technologically-enabled telepathy may carry for the *I* and *other* relationship. If, on the other hand, this technological overcoming cannot be meaningfully accomplished, likewise, we must ask what this fact tells us about the relationship between minds. What is at issue, once again, is to think through what neurotechnologies may tell us about human finitude even in their possible failure to achieve what many commentators claim they will achieve. By thinking through these possibilities (or impossibilities as it may be), it will lead us to a better understanding of the very intersubjective gap that the technologies act upon.

4 TECHNO-TELEPATHY AND THE OTHERNESS OF THE OTHER

> Only later, when I began to probe, did I learn that below the surface transmission—the front of mind stuff which is what I'd originally been picking up—language faded away, and was replaced by universally intelligible thought-forms which far transcended words.
>
> **SALMAN RUSHDIE, *MIDNIGHT'S CHILDREN***

The problem of telepathy, thought-transference, or extra-sensory communication is one that has generally been given a wide berth in serious philosophical or scientific discourse, and relegated to the obscure domains of pseudoscience, mysticism, and new-age quackery.[1] However, while the possibility of telepathy as an innate mental faculty may remain a fanciful one with little scientific credibility, the prospect of direct, technologically-enabled communication between minds is one that is being taken increasingly seriously. Indeed, theoretical physicist and prominent futurologist Freeman Dyson has expressed his belief that what he calls radiotelepathy—"the direct communication of feelings and thoughts from brain to brain"—will be the most significant scientific development of the next eighty years, leading to "radically enlarged" opportunities for shared understanding and thus to our experiencing life "in a whole new way."[2]

The initial steps in this direction have already been taken by scientists combining BCI principles with neural stimulation technology to allow for information to be conveyed directly from one brain to another. Early experiments into brain-to-brain interfacing (BBI) have generated widespread excitement on social media and in the popular press and, perhaps inevitably, reports have surfaced that Facebook is monitoring developments with a view to enhancing the means by which we share thoughts and experiences with friends. As Mark Zuckerberg proposed in a live web chat, "One day, I believe we'll be able to send full rich thoughts to each other directly using technology. You'll just be able to think of something and your friends will immediately be able to experience it too if you'd like."

Imagining Telepathy

In contrast to research in BCI technology, which has developed a more or less standardized set of approaches and aims, where the only real divergences concern the most successful means of recording brain activity, in experimental BBI research there is as yet no such consensus. From surveying the outcomes achieved so far it is clear that there is no single idea of exactly what would constitute communication between brains (or minds), nor is there unanimity on what the purpose or potential future applications of such communication might be. This is no doubt due in part to the emergent nature of the field, which has only begun to take shape in the second decade of the twenty-first century, but it is also due to the huge increase in complexity in the sender–receiver relationship when compared with BCI technology. After all, it is relatively easy to agree on what it would look like to exercise control over a tool or a computer program with the mind, while it is considerably less straightforward to agree on what it would mean to transfer information from one mind/brain to another.[3] Rather than acting on an inert mechanism, such as an electric wheelchair or a computer cursor, in a BBI system the information is being transmitted from one conscious agent to another and this relationship is of a fundamentally different nature to our relationship to an object, as Husserl, Sartre, Levinas, and countless others have shown. As such, neural interfacing between brains introduces a unique set of technical and philosophical problems that do not arise in the interfacing of a brain and a machine.

Some early BBI prototypes have skirted this considerable difficulty by developing an active-passive model of brain-to-brain communication where the second (passive) participant is precisely made into a mere inert mechanism over whose actions the active participant exercises control. Researchers at the Cognition and Cortical Dynamics Laboratory at the University of Washington in Seattle developed a noninvasive human-to-human BBI using EEG signals to harness neural information from the "sender" and transcranial magnetic stimulation (TMS) to transmit information to the "receiver."[4] TMS uses a magnetic coil to deliver electrical pulses to certain specified regions of the brain and has been used as a treatment for severe depression and neurological disorders. The BBI experiment involved two participants in separate locations, one of whom is viewing a rudimentary video game that the second subject cannot see. When the first subject (the sender) intends to activate the cannon fire in the game she engages in right hand motor imagery (imagines moving her right hand) and this sends a signal to the second subject (the receiver) over the Internet, stimulating an area of his brain which causes his hand to jerk and press a touchpad which fires the cannon. Thus the sender's intention is realized through the body of the receiver in much the same way as a standard BCI system induces an activity from a machine. An interspecies version of this experiment was developed by scientists at Harvard Medical School, where brain activity in a human "sender" subject excited the motor area of a rat's brain eliciting tail movement.[5] A 2014 article surveying the ethical implications of brain-to-brain communication even suggests that this latter study may promise future interspecies BBIs working "in the reverse direction, from non-human animals to humans, for such things as enhancing our sensory systems (e.g., improving olfaction by linking our olfactory systems to those of a dog) or aiding in search-and-rescue operations, linking our brains with those of the search-and-rescue animal."[6]

Clearly the human-to-human BBI experiment described above does not "take full advantage of the receiver's capacity for processing information," as the same team of researchers admit in a subsequent paper.[7] Indeed, while it is described in the abstract as a "cooperative visuomotor task" it is somewhat questionable to what degree the receiver could be said to be genuinely "cooperating" in this procedure. To overcome this evident drawback they designed a new model for a BBI system which involved a greater degree of conscious collaboration between the subjects. This experiment, as presented in a 2015 paper, consisted of a question and answer game between two participants, where one subject, the respondent, thinks of an object chosen from a list and the other, the enquirer, asks a set of predetermined questions to discover the identity of the object, eliciting a yes or no response from the respondent.[8] The yes/no choice is detected in the respondent's brain using a standard EEG system such as we have encountered previously, and a signal is transmitted to the brain of the enquirer also using TMS pulses. However, rather than eliciting an involuntary motor response as with the hand movement in the earlier study here the pulses stimulate the visual cortex producing visual disturbances known as "phosphenes," which appear as blobs or flashes of light in the visual field. The appearance of a phosphene in the experiment equated to a yes answer while the experience of a stimulation pulse without the corresponding appearance of a phosphene means no. This same method was used to transmit either a *hola* or a *ciao* message encoded as binary digits akin to Morse code in an earlier study at the Starlab organization in Barcelona. The authors of this latter study claim that the term "mind-to-mind transmission" is more appropriate here than brain-to-brain communication, "because both the origin and the destination of the communication involved the conscious activity of the subjects."[9]

To be sure, we are dealing with very rudimentary signals conveyed at rates far below those of ordinary sensory communication. Given the current limitations of the technologies which all human-to-human neural interfaces to date have employed this is to be expected. Safety guidelines in place for using brain stimulation devices impose strict limits on the strength and frequency of the pulses which can be delivered. However, the outcomes already achieved and the pace at which the technologies are advancing clearly indicate the possibility for more complex forms of communication taking place between brains, even "including the non-invasive direct transmission of emotions and feelings or the possibility of sense synthesis in humans."[10] Already, in BBI experiments with animals where invasive surgery has been conducted to implant neural processors and intracortical stimulation devices directly into the brains of the test subjects, the results have been predictably more dramatic.

Miguel Nicolelis's laboratory at Duke University has been at the forefront of brain-to-brain interface research in animals. An early paper published in 2012 documented the results of a successful trial aimed at establishing remote communication between two rats enabling them to share sensorimotor information to achieve a common task.[11] The motor information was recorded from Rat 1's brain and transmitted to the brain of Rat 2 via intracortical electrical stimulation which induces it to reproduce Rat 1's behavioral choices and thus learn directly from what its partner has experienced. What the findings suggest is that a "direct channel for behavioural information exchange can be established between two animals' brains," effectively amounting to an extended brain (or "brain-dyad"), where cognition is distributed across two organic brains. In subsequent experiments in the same laboratory this same principle was scaled up and expanded to incorporate several brains working cooperatively in what they dubbed a "Brainet." In a 2015 study four rats were connected by BBI and trained to carry out a series of problem-solving tasks which offered rewards when their behavior was synchronized.[12] The results demonstrated that the four-rat Brainet consistently outperformed individual rats carrying out the same tasks. In a separate study, also published in 2015, a Brainet incorporating three monkeys was able to exercise collaborative influence over a BCI-controlled avatar arm, thereby constituting a radically new form of brain-machine interface device in which the "brain" component is made up of several linked brains rather than only one.[13] What is particularly striking about the language used in the written reports is that it is the *Brainet itself* that is treated in each case as the problem-solving agent rather than the individual animals working together as one. The Brainet, they say, "learns to respond to an ICMS [intra-cortical microstimulation] input by *synchronising its output across multiple brains* [my emphasis]."[14] In the eyes of the scientists we are thus no longer talking about a group of individual animals working together but one distributed, multiply embodied super-brain, or "an organic computer." So while current human-to-human BBI systems uphold the active-passive, sender–receiver model of communication, this does not apply to the Brainet or brain-dyad system.

This is a rapidly advancing, highly competitive area of research where practically every published paper presents itself as the "first" to achieve a particular outcome. In fact it resembles nothing less than an arms race, as evidenced by an interview with Andrea Stocco, co-director of the aforementioned Cognition and Cortical Dynamics Laboratory at the University of Washington. When asked for details of future plans and how he intends to realize his stated goal of transmitting complex information such as affective states via BBI, he responds with a guarded "I would rather not say."[15]

Each of the above examples presents a different idea of telepathic communication: from a cooperative sharing or coupling of minds, to the communication of a message via the mind, to a form of remote control over the other's actions. All are familiar science fiction tropes conceived as a means of enabling a more direct form of communication that bypasses the intermediary of spoken, written, or gestural language. While the aims and methodologies vary greatly, each is premised on an overcoming of distance. First, a spatial and temporal distance, which differs only by degree from existing means of telecommunication also designed to enable accelerated information exchange across distances.[16] Second, however, they are designed to traverse *intersubjective* distance: that metaphorical gap or abyss that constitutes our relations with others. In the Brainet and brain-dyad experiments, the goal is an expanded neural network enabling separate brains to communicate with each other in the same way that an individual brain communicates with itself: all rat-subjects become aware simultaneously of what one individual rat has learned. In the human-to-human computer game study the result is the direct and immediate influence over another's motor actions thus causing them to involuntarily carry out the commands of the other, circumventing their will or conscious consent. In the question-and-answer task (and the *ciao* and *hola* task), the aim is to transmit a specific message or item of information from one brain to another. We could also imagine future brain-to-brain communication systems incorporating fMRI cognitive imaging technologies such as Jack Gallant is developing at Berkeley, where an individual's real and imagined visual experiences could be decoded and digitally reconstructed. Here the aim would be nothing less than to *see* what the other *sees* and to experience the "first-person" experience of another.

So given that the singular significance of techno-telepathy does not lie in its speed, convenience, or ability to reduce spatial distance, all of which apply to varying degrees of existing forms of telecommunication, it must lie in this enhanced understanding described by Freeman Dyson above, or the overcoming of the second form of distance. If the advent of techno-telepathy is to amount to a genuine event that could "have a profound impact on the social structure of our civilisation"[17] then it must present us with an unprecedented insight or mode of access that conventional forms of communication are unable to provide. But how is this to be understood phenomenologically, and what does it entail for philosophies of the other? Could such brain or mind-communion, by bypassing the intermediary of language, truly amount to an overcoming of mental solitude and intersubjective distance, allowing us to reach a level of understanding or knowledge of the other that presently escapes us? The central question that will occupy us in this chapter is what exactly it is that we are seeking in this greater understanding that is not already available to us through our regular channels of communication. Again, it will not be a matter of judging the efficacy or capabilities of specific technologies, or extrapolating from these outcomes and speculating upon future developments. Our aim will be to think through the principle of what these technologies aim to achieve, namely, direct communication between minds, and to consider the possibilities and the limits, as well as the implications for intersubjectivity.

We discussed in chapter 3 the impact of the automatic and instantaneous exteriorization of our inner thoughts and desires, bypassing our conscious filters and self-restraint. However, there is much more at stake in the possibility of telepathic insight than the potential embarrassment of having our own most obscene thoughts transmitted for all to see, or the danger of learning things we didn't want to know, such as what our friends "really" think of us. As Žižek notes, with regards to the epistemological problem of "knowing other minds,"

> [If] I were to "really know" the mind of my interlocutor, intersubjectivity proper would disappear; he would lose his subjective status and turn—for me—into a transparent machine. In other words, not-being-knowable-to-others is a crucial feature of subjectivity, of what we mean when we impute to our interlocutors a "mind": you "truly have a mind" only insofar as this is opaque to me.[18]

The question is, if I were to gain technological access to the other's interior space, would this substitute a perfect transparency for the opacity supposedly constitutive of intersubjectivity? Would he or she become knowable in a way that presently escapes me? If so, what exactly would constitute such knowledge? At first glance it seems intuitively self-evident: to *really* know someone's mind in this absolute sense would mean to be given access to that individual's every private thought, desire, sensation, hope, and so forth, to the point at which nothing is held back in reserve. It would mean having so complete an insight into the workings of another person's consciousness that she is no longer *other* at all but known absolutely. But how is this "knowledge" to be communicated or manifested? What form is it to take, and can it ever truly annul the other's primordial otherness? Furthermore, as Derrida asks, where telepathy is concerned "is it even a question here of knowing?"[19]

Žižek's remark has echoes of something Kant says in the second *Critique*:

> Hence one can grant that if it were possible to have so deep an insight into a human being's way of thinking—as this manifests itself through internal as well as external actions—that we would become acquainted with every incentive to actions, even with the slightest, and likewise with all external promptings affecting these incentives, then we could calculate a human being's conduct for the future with certainty, just like any lunar or solar eclipse.[20]

Again, the consequences of gaining this power of insight into other minds would be that they became "transparent machines," whose every thought and action would be as predictable as a wind-up toy. They could never surprise us, lie to us, or keep a secret from us. However, despite this initial agreement between the two statements, there remains a crucial difference. For having said that through gaining this absolute insight we could calculate the other's conduct as surely as any lunar or solar eclipse, Kant goes on to say that "we could nonetheless assert that the human being is free" and not a mere machine:

> For if we were capable also of another view (a view which, to be sure, has not been bestowed upon us at all, but in place of which we have only the rational concept), viz., an intellectual intuition of the same subject, then we would nonetheless become aware that this entire chain of appearances *depends*, with regard to whatever can be of concern to the moral law, *on the spontaneity of the subject* as a thing in itself—a spontaneity for the determination of which no physical explanation can be given at all [my emphasis].[21]

The first thing to note is that this faculty of telepathic insight is a species of intellectual intuition and so it is once more a question of finite as opposed to infinite cognition. Second, is the assertion that even though this intellectual intuition into the mind of the other would make his thoughts and actions exhaustively available and predictable to us, nevertheless something would still be held back which is not reducible to this collection of information and that is precisely his *spontaneity*, upon which the "entire chain of appearances depends." We can know everything that a man will think and do from his birth up until his death and yet on some level he will always escape us and remain outside of our knowing because he is a free, spontaneous being. All of these actions and thoughts, calculable and knowable in advance as they may be, have their source outside of ourselves. In the words of Levinas, "The strangeness of the Other, his very freedom!"[22]

Now this is not strictly speaking the point that Kant intends to make here. Rather, he is concerned with demonstrating that while our behavior is subject to causal determinants we are still free, autonomous beings. Knowing all possible constituent factors influencing a person's behavior still does not abrogate that person of his responsibility, and he can only—and must—take that responsibility *for himself*. The point is that all our behavior can conceivably be ascribed to causal factors—whether external/cultural, such as our upbringing or our environment, or internal/biological, such as our genes or our inherited behavioral patterns—and yet this does not tell the whole story; something will always remain left over in such an account, something indivisible. The upshot is that gaining a complete, comprehensive knowledge of the other's existence, including his interior life, will not get hold of the "person" himself, "refractory to every typology, to every genus, to every characterology, to every classification,"[23] in the words of Levinas. That which constitutes him as a subject is his spontaneity, and this is precisely what can never be grasped as an object of knowledge or placed under the power of *my* spontaneity. Put slightly differently, I could know *what* someone thinks or does, as well as *why* she thinks or acts as she does, but *that* she does it is what escapes me.[24] Hypnotic manipulation or mind control, as in the stories of Dr. Mabuse or the BBI experiment with the remote activation of another's bodily movements, would be the nearest we could get to this power over the "that" of his experience, but at the very moment when I gain this power over his spontaneity he has already eluded me. That which we desire to access ("his very freedom!") will have been lost the moment his actions become my actions, thus as Levinas says of murder, this would be at once both power and impotency.[25]

Incomparably Personal, Infinitely Individual

In popular culture the power of telepathy is routinely portrayed as the ability to "hear" somebody's thoughts as if eavesdropping on one's interior monologue. This is a useful fictional device but it cannot be considered a serious possibility for how to imagine the experience of brain-to-brain communication because it envisages conscious experience as the constant internal narration of one's every thought and experience in full grammatical sentences. A particularly egregious example of this was deployed in the science fiction television series *Heroes*, about a group of ordinary people who began to exhibit supernatural powers, one of whom was endowed with the faculty of telepathy. This ability was counteracted in one absurd scene by the villain opting to "think" in Japanese to disguise his thoughts. However, if I really were able to access the private interior space of another, it could not be so simple a matter as hearing an internal voice (after all, whose voice would this be?) Moreover, if this in fact were the only form that telepathic communication could conceivably take, while it would undoubtedly be radically invasive, it is questionable whether it would enable this absolute knowledge of the other in her innermost self that would divest her of her otherness. In the words of Kierkegaard, the moment I enter language "I express the universal,"[26] so "[seen] as an immediate, no more than sensate and psychic being, the individual is concealed."[27] As Derrida notes, in a reading of Kierkegaard, "[once] I speak I am never and no longer myself, alone and unique."[28] As a result, any truly radical telepathic mode of access that is to be more direct than speech must penetrate beneath the layer of language to reveal the unique kernel of individuality that language conceals.

However, what exactly is this silent core of singular selfhood that is obscured by language? Much of Derrida's early critique of Husserl centers around the latter's reliance on "the existence of a pre-expressive and prelinguistic stratum of sense, which the [phenomenological] reduction must sometimes disclose by excluding the stratum of language."[29] Such an assumption "supposes that, prior to the sign and outside it, excluding any trace and any *différance*, something like consciousness is possible. And that consciousness, before distributing its signs in space and in the world, can gather itself into its presence."[30] On this model language is taken to be a mere accessory, necessary to make ourselves understood to others but not intrinsic to the formation of the thought itself. In which case it would make sense to imagine that when we express ourselves, something of our meaning may be lost in translation and the words chosen are always a form of bricolage, not fitting the thought exactly but approximating to it as best we can with the limited vocabulary available to us. However, as we learned from Croce, a thought is not a thought until it has been articulated. Before its articulation there may be the intention-to-say (internally to ourselves, or externally to others), but this does not constitute a thought that could hypothetically be harnessed technologically and transmitted to another mind who may perhaps be able to understand it with greater clarity than if it were to take the detour through the generality of language. As Wittgenstein writes, "[an] intention is embedded in its situation," and so the intention-to-speak or even intention-to-think before we find the suitable expression for it cannot be abstracted from the language that we speak: "In so far as I do intend the construction of a sentence in advance, that is made possible by the fact that I can speak the language in question."[31] In chapter 1 we asked whether it was possible to imagine that BCMI technology could enable somebody without any prior musical training to compose a piece of music with the mind. This notion, as we saw, conceives of musical ability as something that could be learned subsequently in order to translate an idea that otherwise would be inhibited. However, to again quote Wittgenstein, "one can only say something if one has learned to talk. Therefore in order to *want* to say something one must also have mastered a language."[32] Until I have managed to find expression for the thought even I myself do not know what I mean, for it makes no sense to speak of *meaning* here at all.

This type of articulable thought, which could in principle be communicated from one mind to another, via sensory or nonsensory means, must be distinguished from unconscious, unreflective cognition such as that which takes place when we negotiate a busy high street, for example. If, however, we needed to communicate these skills to a child we would need to formalize and make explicit that which we normally carry out implicitly and unconsciously. In other words, the know-how can only be communicated to others when it is translated into know-that. However, what is truly remarkable about the Brainet technology in development at Duke University is that it enables a sharing of information across multiple sites of cognition *at the implicit, prearticulated level.* A task that could not be carried out by one individual alone can be collectively accomplished without the need for members to consciously reflect on the knowledge acquired in order to share it with others. We could imagine a possible future scenario where parent and child could communicate brain-to-brain via Bluetooth-enabled headsets so that the child can learn to cross the road safely from the parent without the need for any explicit knowledge to be exchanged. The child's brain would learn from the parent's by patching directly onto it. This network or assemblage model of cognition would inevitably render uncertain the boundaries between my cognition and another's. Taken to its extreme it would replace the *I* and *other* relationship with a distributed, nodal form of group cognition as found in certain insect communities, where what matters are the shared goals of the collective rather than those of its constituent members.

Building on this possibility, David Roden suggests that one of the possible consequences of brain-to-brain interface technologies is the realization that "perhaps the much-vaunted privacy of human consciousness is as technology-dependent as our now outdated inability to fly from Paris to New York."[33] Subjective interiority and the separateness from others that this entails would be merely a particular stage of human development consigned to obsolescence by technological advances. If, on the other hand, Roden continues, mental privacy is not a historically specific fact about human experience but pertains to the very essence of that experience which we call human then such insectoid subjects lacking conscious interiority would "no longer qualify as human."[34] Whether human or not, the kernel of selfhood constituted by the privacy and interiority of mind would be dissolved into this "multiply embodied superorganism."[35] Just as with the twenty-first century media forms theorized by Mark Hansen (discussed in chapter 3), Brainet technologies would bypass all subjective participation, operating at the "microtemporal" speeds of neural responsivity which occur well below the threshold of conscious attention. Clearly it would be absurd to suggest that in communicating at this sublinguistic, neural level we gain a better understanding of the other in her singularity than when our exchange takes the detour of language. Rather than accessing some private interiority that the generality of language necessarily conceals, it circumvents individual subjective agency altogether, reconstituting it at the level of the collective.

The question, once again, is where is this secret interiority if it resides neither in language nor in prelinguistic neural activity? If language conceals and direct brain-to-brain patching would destroy the individual qua separate subject, what precisely is the unique core of singularity that an intellectual intuition into the other would reveal?

Nietzsche shares with Kierkegaard the conviction that our true, unique individuality is concealed by language. However, Nietzsche goes further than Kierkegaard and contends that this concealment of singularity by universality extends beyond spoken utterances to our intimate conscious experience itself. According to Nietzsche, that part of ourselves available to empirical apperception is neither private nor personal but inherently common. In a remarkably prescient passage in *The Gay Science* Nietzsche describes consciousness itself in terms surprisingly close to the technologically-mediated hive mind envisaged by Roden.[36] Consciousness, Nietzsche says, is "really only a net of communication between human beings."[37]

In Nietzsche's anthropology, the genesis of conscious self-experience can be traced to mankind's sociality and the need to communicate. On its own account, consciousness is "superfluous," and "[the] whole of life would be possible without, as it were, seeing itself in a mirror." The fact that we are conscious at all can be traced to our vulnerability and lack of self-reliance, as a result of which we are continually dependent upon the assistance of others. In order to be able to enlist others to come to our aid we first needed to formulate a demand to ourselves and become conscious of what our needs were, only then could we make ourselves understood to our neighbors. This nonprimacy of consciousness is true not only of our bodily rhythms and vital functions, but also of our "thinking, feeling, and willing life, however offensive this may sound to older philosophers" (297):

> Man, like every living being, thinks continually without knowing it; the thinking that rises to *consciousness* is only the smallest part of this—the most superficial and worst part—for only this conscious thinking *takes the form of words, which is to say signs of communication,* and this fact uncovers the origin of consciousness. (298–99)

Consciousness is, accordingly, not the atomic nucleus of individuality and the definitive mark of personhood, but a derivative phenomenon, through which one communicates with oneself solely out of the need to communicate with others. What we are conscious of within ourselves, by definition, is only what can be commonly understood. It does not truly belong to "man's individual existence, but rather to his social or herd nature" (299). As a result, since the means by which we know ourselves belongs to the herd and not to ourselves alone,

> [Given] the best will in the world to understand ourselves as individually as possible, "to know ourselves," each of us will always succeed in becoming conscious only of what is not individual but "average." . . . Fundamentally, all our actions are altogether incomparably personal, unique, and infinitely individual; there is no doubt of that. But as soon as we translate them into consciousness *they no longer seem to be.* (299)

What is most *our own* is thus not available to conscious introspection. All access to our own interiority is inescapably mediated by collectivity and so even in our intimate self-experience we know merely the public side of ourselves. This bears a clear resemblance to Derrida's deconstruction of the philosophy of autoaffection, exemplified in his early readings of Husserl. Derrida argues, in short, that there is no immediate reflexive self-experience that is not constitutively contaminated by externality and hence commonality. Where Nietzsche and Derrida part ways, however, is in the former's subsequent assertion of the existence of a preconscious or unconscious self, which is the true, unique individual. So conscious thought may be inseparable from language, but this is only a secondary, derived phenomenon. True thought that is mine and mine alone is structurally inaccessible to me.

We should not avoid remarking upon the obvious here, which is the proto-Freudian character of Nietzsche's thought. In chapter 3 we saw how for Freud word-presentations are that which mediate between an unconscious and a conscious process. By the interposition of word-presentations, "internal thought-processes are made into perceptions,"[38] and by doing so we place ourselves at a distance from them, confronting them as if they were outside of us. In Nietzsche's terms they become no longer our own thought processes but the merely general and average thoughts of the herd. Therefore, in order to make a thought communicable—to ourselves as well as others—we rob it of its truth and betray our innermost selves.

This picture becomes complicated by Freud's claim in "The Unconscious" that communication can take place between one unconscious and another "without passing through [consciousness]."[39] This anticipates his later writings on the subject of telepathy, where he even suggests that "psychical transference" may have been the "original, archaic method of communication between individuals"[40] only later evolving into more advanced forms of communication involving the sense organs (and so before the advent of consciousness according to Nietzsche.) Once again then, brain-to-brain technologies offer support for Freud's claim that technological advances are essentially regressive, bringing about a return to a primordial state. What is more, the moment we introduce the factor of unconscious communication it means that we can no longer ascertain whether an unconscious thought or desire is indeed our own or that of the other. If communication can take place without being rerouted through consciousness, if there is even a minimal communicability to unconscious thoughts and desires, then they must exist in the form of a common or shared "language," otherwise this would be inconceivable. If, then, we are to say with Nietzsche that when a thought becomes conscious it is translated into "the perspective of the herd," following Freud (and brain-to-brain interfaces) this would in fact be nothing more than the translation of one general form into another. For if even our most primordial, unconscious thoughts and desires are marked by commonality and subject to the infiltration of the other, then no matter how far back we regress we will never reach this kernel of individuality or singularity that belongs to me alone and that is not touched by the trace of the herd. Furthermore, were our unconscious innermost thoughts "incomparably personal, unique, and infinitely individual" as Nietzsche contends, how could even the smallest part be transposed into a heterogeneous—conscious—stratum that is common to all? Either the separation between the conscious self and the true, unique (unconscious) self is absolute, making translation from one into the other impossible, or there is already an incipient communicability and universality common to each person's "individual existence." Without this potential there would be no common measure across the different territories, and hence no possibility of communicability—or indeed *nothing but* communicability and no trace of interiority as in Roden's imagined superorganism.

It is clear that the idea of accessing a fully constituted stratum of self (whether one's own or that of the other via telepathy) that exists prior to the "external" recourse to language is untenable. Consequently, since language expresses the general, or herd, perspective we would come no closer to the other in his singularity by means of telepathy than via any other form of communication. However, this still assumes that there *is* a unique singularity that belongs to the other and which language seemingly misses and cannot grasp. But if there is no silent, prelinguistic self, and if language is inherently general, then what exactly *is* there within me that escapes language and is truly my own? It seems we cannot have it both ways: either the other is already intrinsically knowable, because we speak the same language, or there is something unique to her that is incommensurable with linguistic expression and which will always remain outside of my grasp.

Behold Her with My Eyes and She Will Appear a Goddess to You

Even if we are uniquely erudite, we may often find ourselves failed by language when it comes to articulating our most intimate feelings and experiences. Here the inadequacies of expression and its noncoincidence with our private interior space are laid bare. Such experiences are irreconcilable with the generality of words, an obvious example being how people in love very often feel that they are unable to communicate their love to the other and that the word *love* itself is a mere placeholder, barely even approaching that which they wish they could say. This inarticulacy is not an empirical lack, but it is constitutive to the feeling itself that we are unable adequately to communicate it. "How much do you love me?" is a question that we are constitutionally incapable of answering. So there is a structural incommunicability to our most private feelings, leading often to an attempt simply to show the person how it feels by our actions—*this* is how I feel about you; *this* is how much I love you. But if language cannot penetrate the bottomless depths of our experience, what if we could spontaneously exhibit this constellation of emotions in just the manner in which it comes upon us, and transmit it to another mind so that the buzzing cloud of thoughts interlaced with memories that we could never fully outline in words were directly experienced by the other?

In the 1983 science-fiction film *Brainstorm,* a group of scientists are working on a technology that can do precisely what we are describing here: that is, communicate a first person sensory experience directly to another via virtual reality headsets. Their earliest prototype enables the wearer to see what others are seeing, as if experiencing by proxy, or inhabiting the other's body. A subsequent breakthrough allows users to actually transmit the feelings, thoughts, and emotions that accompany and color that experience so that rather than merely seeing through their eyes the others can, as it were, see into their heads. One of the lead investigators on the project, played by Christopher Walken, has become emotionally estranged from his wife over a number of years and communication between the two has completely broken down. As they are on the verge of divorce he "uploads" a series of happy memories of their marriage, along with the emotions that they evoke, and presents it to her as a parting gift. After this seemingly impossible communication takes place they both re-experience their former love for each other and are reconciled. The distance between the two is thus short-circuited, and everything they feel but cannot say to each other can now be directly shown.

Given the results of the cognitive imaging experiments at Jack Gallant's laboratory at Berkeley, it is fair to say that the possibility of transmitting a "mental picture" into the brain of another is no longer solely the preserve of science fiction. However, irrespective of the huge technological challenges involved in reconstructing and transmitting significantly more complex experiences than visual perception, is there any sense to the suggestion that an inarticulable experience such as romantic love could conceivably be communicated to others who do not feel the love that we do? Could we show them by making them see the loved one as we do: see them in their lovableness, as it were? But a feeling of love is not an isolable datum that could (hypothetically) be detached and conveyed to another person. It is an ongoing, mutable network of associations and memories that have developed into the inchoate feeling to which we give the name love.

The same applies not only to a feeling as singular and profound as love but, to a greater or lesser extent, of every thought, sensation, or experience.

Even our most banal and commonplace experiences are intimately bound up with a whole network of prior associations, without their being consciously attended to. This is their meaningful horizon, and one could not detach the "thing" (the momentary thought or feeling) from this context without making of it a different experience. As Wittgenstein asks, "[could] one have a feeling of ardent love or hope for the space of one second—no matter what preceded or followed this second?—What is happening now has significance—in these surroundings. The surroundings give it its importance."[41] If we were to isolate one instant in a person's experience, interrupted from the thread of associations, hopes, expectations, and memories that led up to it and will follow on from it then that momentary "state" would not be the same moment as it is experienced by the person himself. To put the same point in Husserlian terms, the world exists for me "only as *cogitatum* of my changing and, while changing, interconnected *cogitationes*" and hence it follows that it would not be the same *cogitatum* outside of this particular stream of *cogitationes*.[42] The experience of love, as well as the word itself, as Wittgenstein comments, "refers to a phenomenon of human life"[43] and thus has sense only within that life. And, to again juxtapose Wittgenstein with Husserl, the latter expresses something very similar when he writes that "what is experienced" in our moment to moment experience is "nothing more than a synthetic unity inseparable from this life and its potentialities."[44] The whole series of associations that make up my feeling of love toward a person (or of hope, or hatred, or enjoyment) could not be distilled into one communicable flash, just as the whole of a person's life could not be condensed into one second.[45]

To simplify matters we could consider a more tangible example of this experiential complex than romantic love and ask whether we could imagine a technological system that would be capable of isolating and transmitting it to another person's brain. Suppose I am revolted by the taste of bananas: could the experience of someone who does like that taste be transmitted to me, so that I can experience what-it-is-like to enjoy bananas? If we give credence to the philosophical notion of qualia then we would have to assume that we each have a different experience of this taste, and that the distinct qualitative unit of experience for someone who loathes bananas is fundamentally different from that of someone who cannot get enough of them. But there is nothing in the taste of bananas that I am missing, nothing that my bananaphile friend is privy to that I am not. It is simply that we have different taste response characteristics to one and the same stimulus. We could imagine a conversation taking place along the lines of, "I hate the mushiness, the lingering bitter aftertaste," to which my friend replies, "yes, that's what I love about them."

Each individual's network of reactive dispositions is peculiar to her and is constantly adapting and transforming over time as she learns. So does it really make any sense to talk about isolating and communicating individual experiential qualities, as in transmitting the likeableness of bananas to someone who does not like them? All that could conceivably be transmitted in a real-world equivalent of the *Brainstorm* technology would be the stimulus itself. Since there is no transcendent phenomenal property that corresponds to liking or disliking bananas, merely the individual reactions of our taste receptors, I will still not like the taste of bananas even if I experience it by proxy, "through" the experience of someone else who does like the taste. There is no information that could be made available to me that would transform my experience, for we each taste the *same thing*, which is to say the taste manifests itself identically to each of us and we simply have different responses to it. All of this, of course, goes equally in the case of love. Love is not predicated on knowledge, as if, as the Phil Spector song goes, "to know him is to love him." I do not love somebody because I know something about him or her that a third person does not, or because I simply know that person better than anyone else does. The others may know everything that I know about the object of my love and yet still not be able to feel how I do. Thus the quote attributed to Nicomachus with which I have titled this section[46] is not as if to say "if you *knew* her the way I know her you would feel as I do," but is rather the tautological, "if you felt the way I do about her you would feel as I do." There is nothing that the other person is missing in his experience of my beloved that could be made evident to him in a flash of realization.

An even more crucial objection concerning the conceivability of gaining access to another's intimate first-person experience is this: even if we ignore all of the objections raised so far and suppose that we were indeed able to upload our experiential tapestry to a storage device as Christopher Walken's scientist does and enable others to experience it for themselves, it would still not truly be *my* feeling of love that the others experience but their own. Even if it is (originally) "my" network of memories, feelings, and associations that they are able to feel it would still be experienced as if it was their own. She now perhaps knows what it would be like were *she* to feel the love that *I* do and to have had the memories and feelings that *I* have had but what she still does not know (firsthand) is what this experience is like *for me.* And surely if the faculty of telepathy is indeed to effectuate such a radical penetration into the other as to overcome her very otherness then this is the decisive obstacle that would need to be negotiated. It is not an enrichment of my own experience that I would be looking for but an insight into the experience of another.[47]

In Spike Jonze and Charlie Kaufman's 1999 film *Being John Malkovich,* a magical cupboard door exists that leads directly into the mind of the actor John Malkovich. However, the way this is experienced for the successive inhabitants of Malkovich's mind (which is perhaps the only way it could possibly be *experienced* as such) is as if the actor was a puppet or a machine and the inhabitants are the "ghost in the machine," seeing through his eyes and directing his movements but still as themselves, occupying his body in a manner comparable to the BBI experiment in which one subject exercises neural control over the body of the other when playing a computer game. What we are trying to conceive here is whether and how we could really *be* John Malkovich—that is, to not be ourselves experiencing a day in his life. As Derrida discusses, this is the very question of original finitude, or "the *philosophical question in general*":

> [*Why*] is the essential, irreducible, absolutely general and unconditioned form of experience as a venturing forth towards the other still egoity? *Why* is an experience which would not be lived as *my own* (for an ego in general, in the eidetic-transcendental sense of these words) impossible and unthinkable? This unthinkable and impossible are the limits of reason in general. In other words: why finitude, if, as Schelling had said, "egoity is the general principle of finitude"?[48]

The Experience of the Other

In the fifth of his *Cartesian Meditations,* Husserl famously turns his attention to transcendental intersubjectivity, applying the phenomenological method to the experience or constitution of the "alter-ego." Much of the problem centers on how we are to rescue objective knowledge of the external world once we have performed the transcendental reduction, which restricts us to the interiority of our phenomenological presentations. The answer Husserl proposes is that this knowledge is secured through intersubjective agreement: through my knowing that there are other openings onto the world exactly like my own and thus that the world as it is for me is shared by others. However, before we can establish this agreement we must first have encountered these other egos, who are "surely not a mere intending and intended *in me*, merely synthetic unities of possible verification *in me*, but, according to their sense, precisely *others*."[49] The other *qua* other opening onto the world and guarantor of that world cannot be incorporated as merely a moment of my own consciousness without thereby sacrificing that very quality that constitutes him as another ego. The question is how I am to have an experience of something definitively outside of me while still preserving that exteriority.

Throughout this fifth Meditation, Husserl insists upon the insuperable "sphere of ownness" that accompanies all experience, meaning that every thought, every perceived object, every sensation is such only if it is experienced as *my own* experience. This fact of being confined to one's own perspective, which is the condition of all experience, is finitude itself. However, the apparent fall into solipsism that may seem to ensue when we follow the phenomenological method is precisely what Husserl will contest here, for insisting on the insurmountability of one's own experience is by no means to deny intersubjectivity but on the contrary it is to secure it. As Husserl says, "[the] only conceivable manner in which others can have for me the sense and status of existent others, thus and so determined, consists in their being constituted *in me* as others" (128). Any experience of another subject must still be *my* experience if it is to be an experience at all, so any and every encounter with the other must necessarily hold back from the other "himself." In other words, my experience can never reach all the way out. All the knowledge I have of him remains a part of me rather than of him as such, and so in a certain important respect I never truly encounter the other, at least not directly.

This is where Husserl is fundamentally at odds with Levinas, the latter accusing the former of doing violence to the other, and making of him merely an object of my consciousness. For Levinas, the other is not in any respect an alter-ego (in the sense that I view him as another like myself), for it is not a case of starting with my own subjectivity and working outwards. To be sure, Levinas likewise insists that we do not "reach all the way out" and gain access to the other in his innermost being, but in the face-to-face relation with the other he gives me "more than I contain."[50] This relationship, in Levinas's terms, is a *teaching* as opposed to a *maieutics*, the latter merely awakening within me what was already there lying dormant.[51] The other comes to me from a veritable outside; he is presented to me as *absolutely* exterior; "absolute" understood in distinction from the *relative* externality of objects, which, while retaining a certain autonomy, are nevertheless presented *within me* as being outside of me. In the case of the other, however, he is never "within" me in such a way as a thought or representation of mine, for this could only ever reduce him to being something that I contain. So rather than being an object of cognition, the *noema* in the case of the other "overflows the capacity of thought" (49). The only mark or presence of the other within my subjective space is his very transcendence and infinite externality: the other qua other *is* this infinite excess or overflow.

Now in order for there to be this excess there must be a "point of departure" (36) from which it takes flight for the infinite overflowing of thought *needs* the thought that it overflows. Indeed Levinas repeatedly concedes that alterity "is only possible starting from *me*" (40). It follows that the relationship to the other is such only for an *I* that remains itself and does not dissolve in the relationship. The quarrel, then, between these two accounts circulates entirely around *how* the other is presented to this *I*, and how to maintain the other's externality within this presentation. Of course, this was already Husserl's stated problem and by no means something he neglected, but Levinas's contention is that the attempt results in a failure in which the other's *infinite* externality is reduced to the merely qualified externality of objecthood. All of Levinas's beautifully lyrical analyses of the face-to-face relation are oriented around articulating an alternative description that does full justice to the other in his alterity and does not result in objectification.

This relation of the face-to-face is, says Levinas, "totally different from *experience* in the sensible sense of the term, relative and egoist [emphasis added]" (193).[52] The face is "neither seen nor touched—for in visual or tactile sensation the identity of the *I* envelops the alterity of the object, which becomes precisely a content" (194). This is the violence that Levinas ascribes to Husserl's account: the *I* exerting domination over the other by making of him merely a modification of my own consciousness. Again, for Levinas the other is never *given* to me as the content of my thought, "he maintains a relation with me" but "he remains absolute within the relation" (195). The relation is therefore "maintained without violence, in peace with this absolute alterity" (197) and exteriority is never converted into interiority.

We have been trying to consider how we could gain access to the other's ownmost experience without thereby wresting it back into *my* experiential space, however, Levinas seems to suggest that such an approach is inappropriate. Since an experience can only ever belong to *me* it is wrong to speak of an "experience" at all where intersubjectivity is concerned. The other is absolutely irreducible to a subjective event for me. Thus not only do I never experience the other "himself," but I never *experience* the other at all, since the other is never something that happens to me but is something that remains infinitely outside of me. So we can still say that all experience remains *my* experience, but if we follow Levinas we must add the codicil that the other can never be a part or moment of this experience. Returning now to brain-to-brain technology and the distant promise of an immediate exchange between minds, we might ask via Levinas: if it is not *my* experience in the face-to-face encounter, do we already have what we are looking for and hence have no need for telepathic neurotechnologies to accomplish this immediate relation? Is not every encounter with the other already a surpassing or negating of my own experience and a presentation of the other *in person*, that is, immediately?

But this entails further questions, for if my encounter with the other is not an "experience," which is the essence of my sphere of ownness, how am I still to hold onto the *I* and not let it dissolve in the relationship, which Levinas acknowledges to be crucial? Where am "I" in this face-to-face relation? How do we preserve "this radical heterogeneity" (294) unless I hold fast to my experience and the other likewise? If the other is not somehow a part of *my* meaningful existence, if she remains entirely external to any cognition or perception I can have of her, then how is she to present herself to me at all? How can I conceive of the other as such unless she is the other *for me*? Unless the *I* collapses altogether into the other, the only alternative through which she would remain wholly outside of my experience is if I have no encounter with her at all.

Derrida, coming to Husserl's defense, judges that Levinas "deprives himself of the very foundation and possibility" of his own enterprise by "refusing to acknowledge an intentional modification of the ego—which would be a violent and totalitarian act for him."[53] For although Levinas insists upon the "point of departure" in the relationship (the *I*), he cannot account for how the infinitely other appears from this vantage point unless it is as an "intentional modification of the ego"—his appearance in and for me, which, as Husserl shows, is appearance *tout court*. Thus a certain violence is irreducible according to Derrida, and there is no "peaceful" encounter with the other that would not coincide with the worst violence—again, either the total absorption of one into the other or the negation of the face to face encounter altogether. The other can only ever be presented to me as a moment of myself and never as absolutely, or infinitely, other. This violence is preethical and opens the possibility of ethics, in that prior to it there would be no intersubjective relation to speak of. Furthermore, as Derrida here contends, it is only if we follow Husserl and view the other as precisely an alter-ego, "like myself," that we do justice to him in his alterity. For "[if] the other was not recognised as an ego, its entire alterity would collapse."[54]

So it seems we have reached a certain limit, and the only way for the other to appear to me in any kind of relationship ("face to face" or "mind to mind") would be in and through my own experience. Perhaps the telepathic relationship being sought, which would present the other absolutely and without mediation, is the Levinasian ideal of the peaceful encounter. For if I yearn to access the other from her own perspective and not from mine, it is because we surely see this as being the only truly nonviolent relationship: rather than her appearing to me and for me, she would be present for herself, able to *speak for herself* without intermediary. But, to reiterate, this could only be accomplished as the worst form of violence—either to myself or to the other. It is thus only by doing justice to Husserl's account, against Levinas's critique, that we guarantee respect for the other in her otherness. The ability to "depart" from oneself and enter into the other's phenomenological interior absolutely would indeed destroy intersubjectivity, but not by making the other into the absolutely transparent object of my cognition as Žižek describes, but rather by my collapsing entirely into her subjectivity and sacrificing my sphere of ownness. For how could I experience another's phenomenological inner-space, as *she* experiences it, while simultaneously maintaining an external perspective? There is no compromise or median point between these two "points of departure," but the wish is to somehow straddle the divide: to experience her experience while simultaneously maintaining a certain selfhood, which is an impossible hybrid comparable to the Sartrean In-Itself-For-Itself, or God.

So, for Husserl, in our knowledge of the other a "*certain mediacy of intentionality* must be present," which is termed "appresentation." The other is *appresented* but never *given*, for if he were given, "if what belongs to the other's essence were directly accessible, it would be merely a moment of my own essence, and ultimately he himself and I myself would be the same" (109). So if I *were* to gain this impossible access, the very instant it is consummated the other evaporates:

> Whatever can become presented, and evidently verified, *originally*—is something *I* am; or else it belongs to me as peculiarly my own. Whatever, by virtue thereof, is experienced in that founded manner which characterises a primordially unfulfillable experience—an experience that does not give something itself originally but that consistently verifies something indicated—is "other." (114–15)

As soon as I access the subjectivity of the other *immediately* (as soon as he is presented rather than appresented), the other becomes a part of myself and intersubjectivity collapses. So even though telepathy would perhaps bring us to the closest point just short of an immediate encounter, the communicating subjects would still "absolve themselves from the relation"[55] that they enter into, otherwise there would be no relation as such. But our insistence upon the inescapability and the finitude of *ownness* need not equate to solipsism. In fact, as Søren Overgaard stresses, would it not be the other possibility that amounts to the truly solipsistic position? For if I were not condemned to access the world only through my own individual perspective this would condemn me to the "*solitude* of being the owner of *all* perspectives."[56]

From Telepathy to Teleiopoesis

All of the foregoing should allow us to understand the following challenging passage from Derrida's essay "Telepathy":

> For here is my latest paradox, which you alone will understand clearly: it is because there would be telepathy that a postcard can always not arrive at its destination. The ultimate naivety would be to allow oneself to think that Telepathy guarantees a destination that "posts and telecommunications" fail to assure. On the contrary, everything I said about the postcarded structure of the mark . . . is found in the network. This goes for any tele-system—whatever its content, form, or medium.[57]

If we juxtapose this citation with another from the early work *Speech and Phenomena* the consequences for telepathic exchange become even clearer:

> Everything in my speech which is destined to manifest an experience to another must pass by the mediation of its physical side; this irreducible mediation involves every expression in an indicative function.[58]

No matter how clearly one may express oneself, the mediating sign is never transparent. My interlocutor's "lived experience" can never be present to me except by traversing and "to some degree [losing] itself in, the opaqueness of a body."[59] Because of this "irreducible mediation" communication can never be ideal or perfect; there is always a certain interpretative necessity on the part of the receiver and hence the ever-present possibility of failure or misunderstanding. I may hear a perfectly innocent remark as a sarcastic gibe, or I might take literally a merely hollow offer such as a dinner invitation. At first glance, however, telepathic communication would appear to bypass this recourse to externality, promising mind-to-mind communication without mediation. Telepathy then, as a means of "instant" delivery, seems to establish a direct connection between subjects, overcoming the possibility of misunderstanding or miscommunication between the other and myself.

Moreover, not only would the possibility of miscommunication be circumvented in such a direct telepathic connection but likewise the possibility of outright dissimulation would be seemingly annulled. Sartre writes that, "[by] the lie consciousness affirms that it exists by nature as *hidden from the Other*; it utilises for its own profit the ontological duality of myself and myself in the eyes of the Other."[60] But would a telepathic insight not see behind the curtain and perceive the insincerity, thus overcoming (in part) the other's hiddenness? Once again, the television series *Heroes* furnishes us with the perfect example of how *not* to envisage this possibility. In one scene involving the protagonist with telepathic abilities, another character says one thing (out loud) while thinking another, which the telepathic protagonist of course is able to "hear." However, a lie is not like a box that contains something different to what it says on the label. I do not, for example, think no and say yes, I simply *say* that which is not the case or that which I do not really think or believe. As Wittgenstein would say, there is nothing internal in addition to the utterance that accompanies it, except perhaps a feeling of guilt or unease, but then this would be manifest (to the astute observer, or the polygraph) in my behavior. No matter how far we may penetrate the other's consciousness we will always remain on the outside, and always in the position of interpreting "external" behavior. Nowhere will we reach a decisive core at which her meaning or her sincerity may be finally determined. This, of course, is tantamount to saying that mediation remains irreducible.

So all communication, any "tele-system" whatever, must "go via the stars"[61] before reaching (or not) its destination. As long as the two egos are "*separated* by an abyss I cannot actually cross"[62] then no form of absolute communion could ever take place and no communication will ever be ideal. Indeed if this abyss was able to be crossed, or closed altogether, then it would mean the end for one or the other of us and hence the end of communication. Thus we could say in Derridean language that this gap or abyss is both the opening and the limit of communication. Telepathy perhaps brings us to the extreme edge of the abyss, the point at which we come closest to passing over into the other side, but this is also the point at which it presents itself all the more insistently as absolutely unnegotiable. So if anything this abyssal gap between subjects would be more pronounced in the case of telepathy than in sensory communication precisely because by circumventing what *seems* to constitute the obstacle—space, the sense-organs, the constraints of language—and bringing one mind "directly" into contact with another, the very impossibility of realizing this immediacy confronts us—"directly," if it is possible to speak of an immediacy of mediation. As we have intimated above, there would be no possibility of return once we had crossed the limit; or if we did come "back to our senses" there would be no way of integrating this "out of body" (or, rather, "out of mind") experience into our own stream of consciousness without assimilating it and making it into *my own* experience. It would become either a blind spot or a memory of something that happened *to me*, and not a memory of somebody else's experience.

Depending on how we conceive of it then, either there is no telepathy or there is nothing but telepathy. In the former case, understood as immediate communion of minds, telepathy is incoherent. However, understood in the etymological sense of "to touch at a distance," all communication involves a minimum of telepathy: the miraculous fact that I can be understood at all through the opaqueness of the mediating body and you can "know what I am thinking." Brain-to-brain communication would then be a special case of a general telepathy. If the "destination" were guaranteed in advance then there would be no delivery at all and the sender and recipient would be one and the same. So for all communication, a certain distance, and hence a certain telepathy, is necessary.

However, what are we to make of those scattered instances where Derrida speaks of a message that *includes* its destination within it: a letter that "carries its address along and implies in advance, in its very readability, the signature of the addressee"? This is a message which involves, incorporates, and implies its reader or hearer. Yet how can this be so without already guaranteeing its arrival? And after what we have just seen, if the recipient is already assured and fixed in advance how can we even speak of a delivery? As Derrida describes it, "[here] is an arrow whose flight would consist in a return to the bow: fast enough, in sum, never to have left it." So does anything even take place here if the letter is never posted, or the arrow never leaves the bow? Seemingly so, for "[it] will nevertheless have reached us, struck home; it will have taken some time." Traveling at an "infinite or nil speed," this seemingly paradoxical "absolute economy" Derrida labels *teleiopoesis*, in the sense that it "*renders* absolute, perfect, completed, accomplished, finished," while remaining at a "distance at one remove."[63]

There is a certain instantaneity to the teleiopoetic effect, or even better it "advances backwards; it outruns itself by reversing itself" (32). This is a performative structure unlike any other (which Derrida elsewhere calls an "absolute performative"[64]), whereby the announcement or the prediction of the event will already have made it occur. *What* is said or thought happens by virtue of being said or thought. So it is not enough, Derrida writes in "Telepathy," to "foresee or to predict what would happen one day." Rather what is in question is how to think "what would happen by the very fact of being predicted or foreseen, a sort of beautiful apocalypse telescoped, kaleidoscoped, triggered off at that very moment by the precipitation of the announcement itself, consisting precisely in this announcement."[65] Thinking through the effective force of the prediction itself, the "mighty power" of the "might" (exploiting to the full the double meaning of this word, as strength and as uncertainty—"it might happen"), would allow us to "see the difference between *make come* and *let come* vanish at an infinite speed."[66]

This concept of teleiopoesis is developed in *The Politics of Friendship* through a reading of Nietzsche and his call in *Beyond Good and Evil* to a new breed of philosophers of the future. Nietzsche speaks *to* these future philosophers (using first person plural pronouns), and *for* them. He predicts their arrival, announces their presence on the horizon and calls out, beckons, or appeals to them. But, as we now know, the question is not one of the accuracy or verifiability of this prediction but of what is produced, constituted, effected "by the very fact of being predicted or foreseen":

> Nietzsche renews the call; he puts through—from a different place—this teleiopoetic or telephone call to philosophers of a new species. To those of us who

> already are such philosophers, for in saying that he sees them coming, in saying that they are coming, in feigning to record their coming . . . , he is calling, he is asking, in sum, "that they come" in the future. But to be able to say this, from the standpoint of the presumed signer, these new philosophers—from the standpoint of what is being written, from where *we* (Nietzsche and his followers) are writing to one another—must already have arrived. (34–35)

This is an intricate, circuitous structure whose complexity only increases the further we interrogate it. On the one hand, the call announces and constitutes the called: Nietzsche's prophetic apostrophe includes in advance those heirs to whom it is addressed, and delimits the space in which they can appear; the prediction thus brings about its own realization. The moment it is read and understood it will always have been addressed to that very reader who now understands it, but nothing guarantees that anyone *will* understand the call or be there to hear it, which is a first reason why the message can include its reader without thereby ensuring a determined destination. This accounts for the messianic structure that Derrida ascribes to the teleiopoetic effect: when praying for the Messiah to come or wondering when he will arrive I am addressing myself *to* the Messiah, "here and now, to inquire when he will come." In so doing I establish myself as his "herald and precursor" (37), so whenever the Messiah or philosopher of the future does arrive (and nothing guarantees that he will) the call will have *already* addressed him, only him, no matter how much time may have passed between call and response.

On the other hand, however, this is not a simply unilateral movement, with the reader deciding to try and live up to Nietzsche's demand and become the person described or announced in the text. For if someone is able to hear this call then she *already is* the philosopher of the future whom Nietzsche prophesies, otherwise she would fail to understand it, and the appeal would fall on deaf ears. So this recipient does not simply and straightforwardly *succeed* Nietzsche as his disciple. Even as Nietzsche writes his appeal, his interlocutor must "already have arrived," for in speaking *to* this other, the other "precedes," "informs," (42) and "inspires" (41) the sender. Already the order of succession of the teleiopoetic event is unstable: once Nietzsche pens this address the *subsequent* addressees will have already preceded it.

However, there is a further reciprocal exchange at work here, one understood by Borges when he famously wrote that "every writer creates his own precursors."[67] It is not simply that the writer, artist, or philosopher of the future will hear something that nobody else has yet heard (in the writings of Nietzsche for example), for this assumes that the call resounded *prior* to its being heard and was merely waiting for a response. As such the subsequent artist would not truly have *created* his or her precursors but merely been more attentive to them than others. Rather, if we are to do justice to the circuitous path of teleiopoesis, we might say that in hearing the message the recipient effects the sending of it, recalling that strange temporality of Freud's *Nachträglichkeit*, where an event is only retroactively constituted as a "cause" by the "effect." The receipt of the message by the addressee does not leave the sender (and the message itself) untouched: not only is it the *saying* that acts as a "doing," but likewise the *hearing* has a certain performativity.[68]

However, before we risk placing too much emphasis on the hearer or the recipient we must insist again that it is still the case that the addressee is *included* in the message, and that the ones who hear it understand themselves to have been predicted or prophesied. One is answering to a prior call, but only *in* and *by means of* the arrival is it constituted as *already* having been sent. This complex backwards and forwards in time (the before that comes after) also pertains to the identity of the addressee. For, as we have said, even though there must be a certain quality of "already-ness" to the recipients (accounting for the ability to receive the address), the "me" that hears is nevertheless only "already there" *after* I have heard the call. All of this accounts for the backwards travel of the arrow: *the moment it arrives it will already have arrived.* If the call sent out (by Nietzsche or whomever) was known in advance then nothing would transpire in this movement and nothing would truly be heard. I can only hear or receive it, or take it on board, if it is not already known to me. Thus, Derrida speaks in "Telepathy" of a letter that would be,

> launched toward some unknown addressee at the moment of its writing, an addressee *unknown to himself or herself,* if one can say that, and who is *determined,* as you very well know how to be, *on receipt of the letter;* . . . this is quite another thing than the transfer of a message. Its content and its end no longer precede it [emphasis added].[69]

In reading Nietzsche's call to future philosophers and sensing/deciding that it "speaks to me," that he is addressing me, there is of course a certain recognition: a feeling that it awakens something in me that is already familiar. In other words, I must be *hospitable* to it in order to receive the message in the first place. But although I feel this evident familiarity, the thoughts or feelings that it gives rise to within me did not precede the receipt of the message. Only after receiving it do I have the sense that I felt that way all along. For example, if I am deeply affected by the work of a great writer like Proust, and feel as though he speaks directly to my innermost private experience, I can imagine that it was written for me and to me.[70] Proust *telepathically* knew what I am thinking and how I feel. However, if this recognition pertained to something that I simply already knew then reading Proust would hold nothing for me: it would merely be a succession of trite sentiments and truisms. So again, only *after* the receipt of the message will I have *already known* what it has to tell me. However, we must reiterate the complex reciprocity and circuitousness of this relationship, for through my reading the sender and the message itself are altered. This is not only true in the case of a highly accomplished reader of Proust, like Walter Benjamin for example, but even on a banal level, the meaning it has in the context of my life, the significance I bestow upon it cannot have preceded my reading it. Yet *once* I read it, it was already there waiting for me, for the words do not change.[71] As Derrida writes, "[in] this encounter the destiny of a life is knotted, of several lives at the same time."[72] The sender is therefore the "split effect" rather than the "simple origin of teleiopoesis."[73] Prior to the teleiopoetic effect, neither the sender nor the receiver (qua sender and receiver) are fully constituted subjects; they only become the selves that they maintain in the relationship *through* the relationship.[74] All of this, it will be noted, significantly complicates the Levinasian distinction between "teaching" and "maieutics" described above.

We can see now that this magical symbiotic (telepathic) relationship is not limited to such exceptional cases as predicting the coming of the Messiah or the philosopher of the future, or responding to the work of a great artist, but can have general applicability. In a friendship, or in a love affair, both parties—the "me" and the "you"—are to a large extent constituted in and by that very relationship. This is evidenced in a superficial manner by the way in which friends or lovers start to pick up on one another's mannerisms, develop a shared sense of humor, and how we often "behave differently" when we are among different friends.

When I address the other as a friend for the first time, if he or she answers to it in friendship then a certain backwards leap will have taken place: we will already have been friends: "You say 'me' the unique addressee and everything begins between us. Starting out from nothing."[75] Whatever the common ground which preexists our becoming friends (we work together, we attended the same school, we have shared interests), the moment the bond of friendship is made these matters will have been irrelevant. They cannot be the *cause*, since there were others with the same interests, the same experiences whom I do not count as friends, and we can cease to share these interests without thereby ceasing to be friends (indeed with many of the friends we have retained from childhood we may no longer have anything in common other than our friendship itself.) Thus it is founded on a void: the "me" that addresses you as a friend and the "you" that answers are constituted the moment the address is received, for within this relationship neither is who he or she is or was outside of it. Consequently the addressee remains "unknown" even when I *do* know the identity of the person to whom I have spoken, for in answering the address the individual will be other than who he or she was prior to it. Likewise I, as sender, am transformed. Hence this aforementioned leap, from which "everything begins between us." This is again a strangely cyclical movement, where, in the case of the picking up of the other's mannerisms, I imitate you imitating me imitating you, and so on.[76]

This movement is an active passivity and a passive activity that complicates the sender–receiver relationship. We "give by receiving,"[77] and "make come" by "letting come."[78] So we can say that although telepathy/teleiopoesis would not and could not enable me to hurdle the abyss separating me from another, what it does enable is for me to enter into a relationship with the other where we, to a certain extent, become one. We each give birth to something that belongs to both of us, is shared between us, and that neither of us possesses outside of that relationship. As we noted earlier, all of the examples to date of brain-to-brain communication in humans retain and require a direction of travel—a message from the brain of one participant is directly transported to the brain of another. The concept of telepathy developed by Derrida on the other hand denies such an agency between stable identities. It is no longer even a matter of intersubjectivity, which implies a certain occasionalism, and would be a merely external relationship between isolated, fully constituted subjects. However, this cyclical symbiosis remains a closeness in distance, attested to once again by the *tele-* in telepathy/teleiopoesis. I cannot take the other's place, nor he mine, and responding to this address or answering the call, "which you are invited to do to the best of your ability . . . remains your absolutely and irreplaceably singular responsibility."[79] Whether and how we respond can only be assumed by the respondents themselves, there is no program or guarantee.

Responsibility and the Aporia of Ownness

This point about our *irreplaceably singular responsibility* is crucial and brings us back to the passage from Kant's second *Critique* cited near the beginning of the present chapter. There we suggested that freedom, which for Kant is inextricably tied to the moral law, is that which makes my life truly my own. That kernel of individuality that Kierkegaard suggests is concealed by the generality of language is my very spontaneity. This freedom, as freedom from natural determinants, means that nothing can abrogate me from my responsibility, so if I commit a wrongdoing I can have no recourse to mitigating circumstances. I can take responsibility only for myself, and this fact that I am ultimately responsible for my actions and am not a mere causal mechanism is the condition of freedom. So freedom, responsibility and selfhood are all inextricably intertwined. Indeed, as Derrida describes in *The Gift of Death*, through readings of Jan Patočka, Levinas, and Heidegger, responsibility is nothing else than the "experience of one's absolute singularity."[80] Being responsible means experiencing oneself as an irreplaceable individual, in that I cannot pass this responsibility onto anybody else. The consciousness of responsibility is grounded in an experience of death, following Heidegger's (and Patočka's) analysis of death as that which no one can do for me or in my place.[81] It is not as an already established *I* that I take up an attitude or stance toward death, rather the identity of the *I* is bestowed by death: "My irreplaceability is therefore conferred, delivered, 'given,' one can say, by death."[82] Only through this experience of death as *my* death, as uniquely mine, am I constituted as a self that is responsible, and hence free.[83]

As we saw previously, Dasein does not *have* possibilities but *is* its possibilities, and death is its preeminent, "ownmost" possibility in that it can neither be shared with others nor be postponed or avoided. As such it is not one possibility among others but that which is *most my own* and that which opens the very possibility of ownness, experienced as responsibility. It is death that individuates me as uniquely responsible and as absolutely distinct from others. Everything throughout my life that I can do, think, or say could conceivably be done, thought, or said by another. Death, as my own death, is the one "experience" that can be undergone only by each of us alone. So nowhere do I experience my uniqueness and unsubstitutability except in death, in which "all relations to other Dasein are dissolved."[84] Since nobody can die in my place, nobody can ever take over my experience, or know firsthand "what-it-is-like" to be me. If we were not finite beings, burdened with the knowledge that one day we will die, this confinement to one's own experience and perspective would not bear upon us, but this is because there would be no *I* as such.

This then, at base, would constitute the abyss or gap between one subject and another. The impossibility of "really knowing" the other's mind is the impossibility of assuming the other's death. No *unus mundus*, no telepathy could ever overcome this gap without destroying us along with it. For what each of these concepts affirm is the possibility of a *shared* knowledge or experience, and this indeed puts into question the privacy of any specific experience and the extent to which any of my thoughts, ideas, opinions, and affective sentiments could be said to be decidedly my own: in all of these aspects my place can be taken by another. However, death, responsibility, and freedom constitute the point at which my experience becomes my own, that is, uniquely singular and distinct from that of the other.

Insofar as nobody can take responsibility for me (as in take over my responsibility) or die for me (in the sense of dying in my place, taking my death away from me) then each person's experience is indeed, as Nietzsche holds, "incomparably personal, unique, and infinitely individual." First, since responsibility (qua experience of selfhood) is always mine alone it cannot be compared with anyone else's. We may speak of collective or shared responsibility, but in the final instance it remains up to me to take up that shared responsibility. As Kierkegaard insists, I cannot appeal to general or universally applicable laws to guide or justify my behavior, for this would be an abrogation of my own singular responsibility. Levinas, paraphrasing a character from Dostoyevsky's *The Brothers Karamazov*, often writes, "All men are responsible for one another and I more than anyone else."[85] Thus my responsibility is incomparable because I am *more* responsible than the other: *my* responsibility is all that I can speak about and so it cannot be compared or weighed up against another's. As soon as I factor in the other's responsibility (toward me for example), I disavow my own responsibility toward him, introducing calculation and scheming into my behavior. Furthermore, "[how] can I admit his responsibility for me without immediately finding myself . . . responsible for his responsibility itself. To be me is always to have one more responsibility."[86] It follows that this excess of my responsibility over the other's is an exponential escalation, "*increasing in the measure that it is assumed,*"[87] which accounts for the "infinity" of responsibility. For since responsibility *is* the experience of selfhood itself, there is no point at which I will have been done with it, or discharged myself of it. The moment I am no longer responsible, I am no longer a subject.

However, it will not have escaped notice that there has been a slight but fundamental shift in our account of the uniqueness of responsibility in the last paragraph. From resting on an essential solitude (the solitude of death), it has now taken on an evident sociality. Responsibility for oneself, for one's own actions and behavior, has passed into responsibility for, and toward, the other. Levinas never ceases to reproach Heidegger for privileging one's own mortality ahead of all others and insisting upon our necessary separation-in-death. Levinas even suggests that selfhood, or ipseity, derives not from my own death at all but rather from the other's death. The upsurge of responsibility that ensues from my experience of the other's mortality precedes consciousness or free commitment. This is a responsibility even before there is any *I* able to take up or assume that responsibility. It is a responsibility, finally, which did not arise in or from a decision taken by a free subject, but which originates outside of me and assigns me to be a subject. For Levinas, one's singularity as *I* arises through one's being "chosen without choosing,"[88] which is a passivity infinitely more passive than receptivity, since receptivity presupposes a capacity or ability to receive that precedes that which it receives. Consciousness, as Levinas often writes, is inseparable from *activity*: anything that affects consciousness from the outside can always, by its assenting to it, be assumed by consciousness as if it were its own invention. Responsibility to the other, however, cannot be something freely decided upon or assumed by a fully constituted subject, nor can it be something empirically learned through the experience of sharing the world with others. I cannot decide to be or not to be responsible for the other, it is something that falls upon me prior to all decision: "In the face of the other man I am inescapably responsible and consequently the unique and chosen one."[89] As such, rather than coinciding with freedom, for Levinas responsibility *precedes* my freedom, and even precedes the *I*, in an "immemorial past"[90] that cannot be recovered or brought under my spontaneity. So if this hyperbolical responsibility to the other precedes the *I* and calls upon me to be an *I* then we can perhaps understand how "the death of the other preoccupies the human *being-there* [*Dasein*] before his own death."[91] Before taking up a position toward my own death I am concerned about, and implicated in, the other's death. Henceforth, death can no longer be thought of as that which annuls or unties all relations to the other and brings me before myself in my "ownmost nonrelational possibility," but as primarily that which divests me of concern for myself and my own being.

According to Levinas, then, it is the other's death that is preeminent, and it is concern over the death of the other, over his mortality and vulnerability, that calls me to ipseity:

> This would be the *I* of the one who is chosen to answer for his fellowman and is *thus* identical to itself, and *thus* the self. A uniqueness of chosenness! . . . The priority of the other over the *I*, by which the human *being-there* is chosen and unique, is precisely the latter's response to the nakedness of the face and its mortality. It is there that the concern for the other's death is realised, and that "dying for him" "dying his death" takes priority over "authentic" death.[92]

Thus it is still in responsibility, as that which concerns only me, that I experience my irreplaceability and uniqueness but this irreplaceability is awakened in answering to the call of responsibility to the other rather than in facing up to my own mortality. Only *I* can answer to it; he addresses me, and I cannot defer or avoid this responsibility for even fleeing from it or disavowing it is still responding to it. This is an *I* that is elected or chosen, born outside of itself, in sociality, rather than an *I* that arises by bracketing off social relations in its ownmost possibility. The important point is that for Levinas, contra Heidegger, the quality of "ownness" characterizing selfhood is derived and secondary, and *the other precedes the I.*

It obviously cannot be a matter of deciding between one or the other of these two positions, or proving or disproving one or the other. But whichever death precedes the other in terms of priority, whether this is an ethical or ontological priority, nothing Levinas can pose in objection to Heidegger can deny the latter's key insight that my concern for the other's death, even if this takes primacy over concern for my own death, cannot lead to my dying that death for her, and as such taking it away from her. Likewise, this priority of the other's death over my own cannot alter the existential fact that it is in every case *my* death that I will undergo, and never the other's. Even if I took a bullet for another, though I may have saved him from dying at this particular juncture and thus prolonged his life, I have in no way taken his death away from him or died in his place. Nor have I died or experienced *his* death (meaning that the death that I undergo is not mine but the other's) even if this death (the bullet that kills me) was meant for someone else. And as Derrida writes,

> Once it is established that I cannot die *for* another (in his place) although I can die *for* him (by sacrificing myself for him or dying before his eyes), my own death becomes this irreplaceability that I must assume if I wish to have access to what is absolutely mine.[93]

Interestingly, the film *Brainstorm*, discussed earlier, approaches this question of whether one's own death could be taken over or undergone by the other. In the film, one of the researchers has a fatal cardiac arrest in the laboratory while wearing the headset device and thus "uploads" her death for Christopher Walken's character to experience. Unfortunately at this point the film lapses into ponderous metaphysical fantasy involving some kind of celestial afterlife, and by thus conceiving of death as a passage into another place, and hence as an experience—the *ultimate* experience, in every sense of the word—the real problem is avoided, and the true "experience" of death as one's "*ownmost nonrelational possibility not to be bypassed*," that is already presupposed and "understood" in any such metaphysico-theologico-fictional speculation is missed. Admittedly however, it is difficult to conceive of a different route for the film to have taken at this point, since "death" cannot be represented without dissimulating it as a phenomenon of life. But then, if death cannot be presented, made known, or "experienced" *at all*, how am I to have this immediate encounter with my death, as my most proper possibility, that individuates me as an irreplaceable subject? This is the problem Derrida raises in *Aporias*: if death is, as Heidegger often phrases it, the possibility of the impossibility of Dasein—the possibility of Dasein's "no-longer-being-able-to-be-there"[94]—would this be possibility as impossibility, or impossibility as possibility? The two alternatives are not interchangeable, for while the former cancels or annuls itself as possibility (possibility would be precisely that which is no longer possible), the latter still appears within the space of the possible: impossibility would be able to appear as such, as something Dasein is capable of.

As Derrida stresses, Heidegger's whole existential analytic rests on the latter interpretation, on death being a possibility-of-being that presents itself and "*to which Dasein relates*."[95] Death, as the very impossibility of all presentation, all appearing—"the disappearance, the end, the annihilation of the *as such*"[96]—must, for Heidegger's analysis to hold, nevertheless present itself to Dasein—"as such." Even though it is impossibility itself, it must be awaited or anticipated *as possibility*: indeed as our "ownmost possibility." It is this experience of death—the paradoxical appearance or presentation of the impossibility of all appearance or presentation—that distinguishes Dasein from all other forms of life. But if we can no longer appeal to such an experience how is this distinction to be upheld? For if "the impossibility of the 'as such' is indeed the impossibility of the 'as such,' it is also what cannot appear as such" (75). And if animals do not have an experience of, or relation to death as such, "neither does man, that is precisely the point!" (76).

It follows that perhaps the somewhat clumsy and unsophisticated account of the experience of death presented in the film *Brainstorm* cannot be so clearly distinguished from the authentic, primordial understanding described by Heidegger. While the former resorts to a metaphysics of death, speculating upon what lies beyond, and thus betraying the prior, existential experience of mortality, the latter likewise relies on death being something that can be presented "as such": on impossibility appearing as possibility. The demarcation between the metaphysical understanding of death and the more originary or fundamental experience is thus "untenable," in that "it cannot even claim to have any coherence or rigorous specificity" (79). For what could this experience of death *be* that Heidegger opposes to the inauthentic one? How could we "relate" to that to which by definition we cannot relate?

So death, which must be assumed or taken up as our most proper possibility, is also, at the same time "*the least proper*" (71), and that to which I *must* relate as what is most my own is also that which is least my own, to which I *cannot* relate. This knot cannot be untied by taking Derrida to mean that Heidegger is wrong in his insistence on the "mineness" of death, for it remains the case that each dies her own death, and that as such it is surely my death that is most properly mine and no one else's. However, since I cannot relate to my death, it is simultaneously that which is *least* my own and which, in fact, has nothing to do with me. The most and the least, the proper and the improper coincide and cannot be resolved into the one or the other. There is an original contamination in the space of "my ownness." This relation without relation accounts for how we can have such an irresolvable conflict surrounding death as that between Heidegger and Levinas, for both rely on an experience of death *as such*—whether my own or that of the other—that is prior to any empirical knowledge or experience of somebody actually dying. This regression to the most authentic, or the most original, assumes an a priori, grounding encounter with death as such. However, if there is no such encounter, if we can have no relation with death, then the question as to whose death is primary must remain undecidable. Not only can I never experience the other's death immediately, so I can never have an experience of my own death either. Whether it is the other's death or my own death that individuates me as an irreplaceable subject, it is in either case inaccessible so the question remains unanswerable. Just as Heidegger unravels Kant's attempt to establish the stable foundations of finitude (in grounding sensibility and understanding in the free act of the imagination), so here Derrida does the same to Heidegger. The foundational experience of death that grounds mineness and the whole existential analytic of Dasein is swept away, but not by passing to an even more "radical, originary or fundamental thought" (79), but by denying the possibility of any such foundational principle.

The important point to take from this is that there can be no grounding experience of selfhood or *ownness* that is not originally impure. The implications for our discussion surrounding telepathy and the other are clear: at precisely the point at which we are most separated from the other, and closest to ourselves, we are furthest away from ourselves. From the beginning of this chapter, in trying to think whether "I" can ever attain to an immediate experience of the other's experience, we have repeatedly tried and failed to trace a clear line of separation between that which is one's own and that which also belongs to the other. While the original communicability, and hence universality, of every thought and experience was conceded, we seemed to have reached the ultimate point of absolute uniqueness and singularity in death. Here, at least, was an "experience" that was truly mine and which could not be expropriated, but on further investigation it turned out to have been the very essence of expropriation. So if mineness cannot be rigorously grounded or delineated, neither, it seems, can my singularity as distinct from universality. Of course this is not to say that I would ever confuse myself with anyone else, or dissolve altogether into generality, but any attempt to establish what it is that distinguishes me from all others will always be frustrated. Just as it cannot be stated whether it is my own death or that of the other which takes precedence, so it cannot be decided whether intersubjectivity or individual, monadic subjectivity comes "first." When setting out from the presupposition of intersubjectivity, there seems to be *nothing but* intersubjectivity and when positing individual subjectivity as point of departure, we inevitably have to negotiate the problem of solipsism. If this is a necessary undecidability then it cannot simply be the case that there is *nothing* that is truly my own, which would mean there is nothing outside the universal. However, nor could it be the case that there is *something* that is truly my own: some isolable, founding property.

It is this irresolvable, quivering tension that has lain at the root of all of our questions in this chapter. Because the universal does not reach all the way down this encourages the idea that there is some nucleus of otherness or uniqueness, outside of generality, that perhaps an absolute telepathic insight (or intellectual intuition) would enable us to access. This theme recurred repeatedly: language seems necessarily to conceal some secret kernel of selfhood that we each possess individually. However, such a supposition leads to the impasse that we found in Nietzsche, where the two sides of the self could not be reconciled. There we suggested that either there is an inherent communicability and hence universality already belonging to this true, unique self, which undoes it from within, or there is a total absence of communication between the inner self and the "herd" self, which leaves us with a core of selfhood that is both most our own and least our own, which has everything and nothing to do with us. This would be a "self" hidden even from him or her to whom it supposedly "belongs."

For Derrida, such an irreducible, absolutely private self is an inescapably theological assumption, necessarily presupposing the infinite insight of God to whom it would reveal itself. In fact Derrida goes further than this and even suggests that "God is the name of the possibility I have of keeping a secret that is visible from the interior but not from the exterior."[97] In order to assume this inaccessible interiority we must coposit a place or point of view from which it would be visible, as well as a witness to occupy that position. This witness that is so constituted is "*at the same time other than me and more intimate with me than myself.*"[98] So in wishing to gain access to this singular otherness of the other, we are once again striving after the absolute perspective of God, but this is a point of view that is coextensive with the object viewed. In other words, it is a God constituted in order to secure the possibility of this interiority, with each necessarily implying the other. Rather than a transcendent being able to see all and know all, God *is* this radical interiority that would allow each of us to withdraw from the universal. This points toward a different interpretation of intellectual intuition, for if the latter is a cognition whose object, rather than preexisting it, is originally created in and through that cognition, then here also we have an "object" (this secret interior self) that only *is* through divine observation, and which can *be* only so long as it is being observed. However, here the relationship is reversed, for it is effectively the *object* that creates its observer. Only so long as we assume the object itself can (and must) there be this witness to it. If, as we suggested above, there is *no* property of ownness, then the absolute point of view vanishes with it.

So long as there is no unique, singular kernel of selfhood, this desire to see into the other in his or her core of otherness—to gain access to the vantage point of God—will always be frustrated; we will always be on the outside, in the universal, but at the same time knowing that the universal does not account for everything. This, however, is not due to our being unequal to the divine gaze of the Creator, or being incapable of penetrating deeply enough; rather, *the very thing that would ensure the unknowability of the other would be that there is nothing (or no thing) there to know.*

Otherwise than Finitude or Finitude Thought Otherwise?

It remains for us to try and reconcile this outcome with that reached at the close of the previous chapter. There, in our discussion of Jung and our adopting of the concept of *unus mundus*, we appeared to abandon caution in favor of a more adventurous, speculative approach, breaking with the theme of a melancholy finitude which keeps us forever at one remove from the real and embracing instead a monist metaphysics where the gap between internal and external is overcome. So where does this final chapter, which appears more deflationary in its conclusions, sit in relation to what has been argued in the pages immediately preceding it?

What Jung affirms under the name of synchronicity is that there exists a form of intuited content (or "'immediacy' of psychic images") that is neither "received" nor "produced" by the subject. The event of synchronicity is the acausal co-occurrence of separate manifestations—psychical and physical—of the "same living reality." Could a similar statement, with certain precautions, not be made about the form of telepathy that we have arrived at? For we have suggested, via Derrida, that telepathy is not that which will enable me to break out of my own cloistered interior and gain absolute access to the private inner space of another, but is rather that which renders such interiority suspect, necessitating that we reconceive the relationship altogether. This was what lay behind the move from *telepathy* as it is commonly envisaged to *teleiopoesis*, which can no longer be considered along the lines of the simple transmission of a message between two or more self-identical subjects; rather, each is reconstituted in and by the liaison. Like synchronicity, this is no longer a question of *activity and passivity* but of a quasi-mysterious sharing or sympathy. So when we described the relationship as a form of symbiosis in which each subject gives birth to something that none possess individually we could perhaps consider this, too, as being the *respective manifestations*—in me and in the other(s)—*of the "same living reality."* There is thus no stable dividing line between me and the other, for telepathy (or teleiopoesis) undermines the very notion of a fully constituted, separate, self-contained subject. Even death, that ultimate punctuation point, cannot be the locus at which this relationship is stabilized and the identity of the egos firmly fixed and isolated.

Admittedly, the space for convergence between Derrida and Jung here is small and contestable, and it has involved a certain bending of terminology. For this *same* (living reality) is no doubt not *the same* in each case. While synchronicity involves a bifurcation of what is essentially *one*, there is no "essence" or "identity" of what takes place in telepathy/teleiopoesis that is (subsequently or simultaneously) instantiated in separate subjects, as if it were a self-identical, transpersonal content. We have already noted that the *unus mundus* hypothesis, while offering a way to reconceive the relation between thought and external reality, has little to say regarding the intersubjective relation. Thus the final question at the end of chapter 3, which led into our discussions around telepathy, asked: if the gap between *thought* and *that which is given to thought* no longer holds sway, what about the multiple "sites" or "instances" of thought? Does it follow that we are all in some sense "the same" on this antecedent substratum that is prior to the differentiation of thought and being? Yet this would to see all singularity lost in universality. While *unus mundus* posits a certain shared, universal structure of experience we must also be able to account for that which withdraws from universality if we are not to dissolve into an undifferentiated oneness: the famous night in which all cows are black. This is why the outcome of the discussion on telepathy has been rather more ambiguous than that reached in chapter 3, for this lack of a stable point of separation between subjects and the reduction of distance does not amount to a neutral unity. Rather, the gap here is simultaneously *reaffirmed as it is undermined.* Applying a familiar locution of Derrida's[99] we might say that *there is* mineness, interiority, separation, but that this is not a limit that can be touched or delineated. That which individuates me and keeps me at an infinite distance from the other is not a thing, a quality, a property, or an "experience." However, this negative claim must be taken further and turned around into a positive statement: *it is the very fact that there is no such thing, quality, property, or experience that ensures the finite gap of unknowability between subjects.*

Finitude thus returns, but it is no longer a negatively defined finitude of lack or privation, positing a reality that we are forever separated from.[100] Nor is it characterized in opposition to a limit concept, say of intellectual intuition, in which (or in whom) this lack is overcome. For the very idea that there is something *more*, something *private*, that telepathy would enable us to gain access to, is precisely what must be abandoned.

In each case there was something that presented itself as ineluctably beyond reach: a *great outdoors* separated from a closed interior, and each time we have sought to rethink that relation, albeit in different ways. In the first instance this was accomplished by annulling the separation itself and in the second by refusing the very idea that there is "something" that we are separated *from* or that we could reach: something positive and self-identical. Only as long as we dream of overcoming finitude via technology does it present itself to us as *in need of being overcome*. We saw that the idea of an irreducibly secret self necessarily assumes the existence of a divine observer who can see what we cannot. Likewise, the idea that we are ultimately divorced from reality, as Meillassoux has shown, not only opens a space for but actively demands the thought of a deity to fill that gap.[101] By thus denying the gap, in its different ways, we are not attaining to the position of the divine observer and by doing so attaining the faculty of intellectual intuition, but on the contrary refusing the possibility of any such infinite vantage point or faculty.

NOTES

Introduction

1. Cited in Anthony Cuthbertson, "Ray Kurzweil: Human Brains Could be Connected to the Cloud by 2030," *International Business Times* (2015), http://www.ibtimes.co.uk/.
2. Hannah Arendt, *The Human Condition*, 2nd ed. (Chicago, Ill.: University of Chicago Press, 1996).
3. Friedrich Willhelm Joseph von Schelling, *System of Transcendental Idealism* (1800), trans. Peter Heath (Charlottesville: University Press of Virginia, 1978).
4. This is a veiled reference to Transhumanism, the cultural and intellectual movement dedicated to advancing the cause of human enhancement through biological and technological interventions. In continuity with the tradition of eighteenth century enlightenment humanism, transhumanists affirm and celebrate the self-fashioning autonomy of the human species. Their credo is most vividly expressed in the second point of the most recent version of the "transhumanist declaration": "We believe that humanity's potential is still mostly unrealized. There are possible scenarios that lead to wonderful and exceedingly worthwhile enhanced human conditions." Transhumanists consider it an ethical duty to invest research efforts into the enhancement opportunities afforded by new biotechnologies with a view to ultimately advancing ourselves to a posthuman stage in which biological constraints (chief among them, death) have been transcended and finitude definitively surpassed. In unabashedly teleological terms, this is conceptualized as the completion of a process, "like the full moon that follows a waxing crescent, or like the flower that follows a seed." Nick Bostrom, "Letter from Utopia," *Studies in Ethics, Law and Technology* 2, no. 1 (2008): 1–7. See also, Ramez Naam, *More Than Human: Embracing the Promise of Biological Enhancement* (New York: Broadway Books, 2005); Ray Kurzweil, *The Singularity is Near: When Humans Transcend Biology* (New York: Penguin, 2006); Julian Savulescu and Nick Bostrom, eds. *Human Enhancement,* (Oxford: Oxford University Press, 2009); Max More and Natasha Vita-More, eds. *The Transhumanist Reader: Classical and Contemporary Essays on the Science, Technology, and Philosophy of the Human Future* (Hoboken, N.J.: John Wiley & Sons, 2013).

5. See the 2013 report published by the Nuffield Council on Bioethics, "Novel Neurotechnologies: Intervening in the Brain," available online at http://nuffieldbioethics.org/project/neurotechnology/.
6. Douglas Kahn, *Earth Sound Earth Signal: Energies and Earth Magnitude in the Arts* (Berkeley, Calif.: University of California Press, 2013). A CBS news report from inside Dewan's laboratory is available on YouTube.
7. B. Graimann, B. Allison, and G. Pfurtscheller, "Brain Computer Interfaces: A Gentle Introduction," in *Brain Computer Interfaces: Revolutionizing Human-Computer Interaction,* ed. B. Graimann and B. Z. Allison (New York: Springer, 2010).
8. Jacques J. Vidal, "Toward Direct Brain-Computer Communication," *Annual Review of Biophysics and Bioengineering* 2 (1973).
9. Luca Citi et al., "P300-Based BCI Mouse with Genetically-optimized Analogue Control," *IEEE Transactions on Neural Systems and Rehabilitation Engineering* 16, no. 1 (2008).
10. Femke Nijboer et al., "A P300-based Brain-Computer Interface for People with Amyotrophic Lateral Sclerosis," *Clinical Neurophysiology* 119 no. 8 (2008): 1909–16.
11. From the press release at http://neurosky.com.
12. Meel Velliste et al., "Cortical Control of a Prosthetic Arm for Self-Feeding," *Nature* 453, no. 7198 (2008).
13. https://www.youtube.com/watch?v=Y6fug4pzU4Q.
14. Diana Lutz, "Epidural Electrocorticography May Finally Allow Enduring Control of a Prosthetic or Paralyzed Arm by Thought Alone," https://source.wustl.edu.
15. Thomas Naselaris et al., "Bayesian Reconstruction of Natural Images from Human Brain Activity," *Neuron* 63 no. 6 (2009): 902–15; Shinji Nishimoto et al., "Reconstructing Visual Experiences from Brain Activity Evoked by Natural Movies," *Current Biology* 21, no. 19 (2011): 1641–46. Fascinating video footage of Gallant's experiments is available on YouTube.
16. Patrick Tucker, "Could a Brain Scan Protect U.S. Troops from Insider Attacks?", http://www.defenseone.com.
17. Slavoj Žižek, *The Parallax View* (Cambridge, Mass.: The MIT Press, 2006), 193.
18. Theodor Adorno, *Minima Moralia*, trans. E. F. N. Jephcott (London: Verso, 2005), 81.
19. Ibid., 80.
20. Thomas Metzinger, *Being No One: The Self-Model Theory of Subjectivity* (Cambridge, Mass.: The MIT Press, 2003), 314.
21. Metzinger goes on to say that it is a "phenomenological fallacy" to imagine that this is a necessary feature of *all* conscious experience and cites a series of neurophenomenological case studies to make the point.
22. Immanuel Kant, *Critique of Practical Reason*, trans. Werner S. Pluhar (Indianapolis, Ind.: Hackett, 2002), 147. All references to Kant's texts are given by the *Akademie* edition pagination, except references to the *Critique of Pure Reason*, which follow the standard convention of indicating "A" and "B" for the first and second editions.

1. The Idea Becomes a Machine That Makes the Art

1. Title taken from Sol LeWitt, "Paragraphs on Conceptual Art," in *Art in Theory, 1900–2000: An Anthology of Changing Ideas*, ed. Charles Harrison and Paul Wood, 2nd ed. (Malden, Mass.: Blackwell Publishers, 2003), 834.
2. Ibid., 839.
3. Ibid., 836.
4. Lucy Lippard, *Six Years: The Dematerialization of the Art Object from 1966 to 1972* (Berkeley: University of California Press, 1997), 40.
5. Ibid., 157.
6. In particular, Peter Osborne in *Anywhere or Not at All: The Philosophy of Contemporary Art* (London: Verso, 2013) argues that contemporary art, as post-conceptual art, emerged from the necessary failure of first-generation conceptual artists to rid art of its aesthetic, material dimension in a pure conceptuality. For Osborne, all art operating in the wake of conceptualism must have a conceptual, anti-aesthetic component but this is no longer considered sufficient. Rather, this anti-aestheticism can only be worked out through the use of aesthetic materials. See also: Luke Skrebowski, "Conceptual Aesthetics" in *A Book About Collecting and Exhibiting Conceptual Art after Conceptual Art*, ed. Sabeth Buchmann, et al., (Koln: Verlag der Buchhandlung Walther Konig, 2013), and Sebastian Egenhofer, "Aesthetic Materiality in Conceptualism" in *Aesthetics and Contemporary Art*, ed. Armen Avanessian and Luke Skrebowski (Berlin: Sternberg Press, 2012), and, from an analytic standpoint, James Shelley's "The Problem of Non-perceptual Art," *The British Journal of Aesthetics* 43 no. 4 (2003).
7. John Dewey, *Art as Experience* (New York: Pedigree, 2005), 67.
8. In this context the work of John Roberts on the nature of artistic labor after the progressive deskilling of modernity that culminated in Duchamp's "unassisted readymades" is absolutely crucial. See John Roberts, *The Intangibilities of Form: Skill and Deskilling in Art after the Readymade* (London: Verso, 2007); "Art after Deskilling," *Historical Materialism* 18, no. 2 (2010).
9. Immanuel Kant, *Anthropology from a Pragmatic Point of View*, trans. Victor Lyle Dowdell, revised and updated by Hans H. Rudnick (Carbondale: Southern Illinois University Press, 1996), 173.
10. Samuel Beckett, *Three Novels: Molloy, Malone Dies, The Unnameable* (New York: Grove Press, 2009), 29–30. Whether intentionally or not this is almost a direct quotation of Hegel: "Speech and work are outer expressions in which the individual no longer keeps and possesses himself within himself, but lets the inner get completely outside of him, leaving it to the mercy of something other than himself. For that reason we can say with equal truth that these expressions express the inner too much, as that they do so too little: too much, because the inner itself breaks out in them and there remains no antithesis between them and it; they give not merely an expression of the inner, but directly the inner itself; too little, because in speech and action the inner turns itself into something

else, thus putting itself at the mercy of the element of change, which twists the spoken word and the accomplished act into meaning something else than they are in and for themselves, as actions of this particular individual." In the words I speak or the work I produce I estrange myself from myself. The object is not an external sign pointing back to something inner that remains mine, but *is* my inner intention externalized—I do not keep anything back in reserve. But at the same time, since I am powerless to determine its reception, what is intended may become lost or altered in transmission. So it would seem that I am fully present *and* wholly absent from my product or utterance. Georg Wilhelm Friedrich Hegel, *Phenomenology of Spirit,* trans. A.V. Miller, (Oxford: Oxford University Press, 1977), 187.

11. James Lord, *A Giacometti Portrait* (New York: Museum of Modern Art, 1965), 10.
12. This echoes what Levinas has written about *indolence,* as "the impossibility of beginning": "To begin is to begin in the inalienable possession of oneself. It is then to be unable to turn back; it is to set sail and cut the moorings. From then on one has to run through the adventure to its end. To interrupt what was really begun is to end it in failure, and not to abolish the beginning. The failure is part of the adventure. What was interrupted does not sink into nothingness like a game. This means that an action is an inscription in being. And indolence, as a recoil before action, is a hesitation before existence, an indolence about existing." Emmanuel Levinas, *Existence and Existents* (Pittsburgh, Pa.: Duquesne University Press, 2001), 15. Much of Beckett's work, and the *Trilogy* in particular, is the experience of this "indolence about existing."
13. John Barth, *The Sot-Weed Factor* (London: Atlantic, 2002), 415.
14. Michel Foucault, *History of Madness,* trans. Jonathan Murphy and Jean Khalfa (Abingdon: Routledge, 2006), 536.
15. Jacques Derrida, "Parergon," in his *The Truth in Painting,* trans. Geoff Bennington and Ian Mcleod (Chicago, Ill.: University of Chicago Press, 1987).
16. Immanuel Kant, *Critique of Pure Reason,* trans. Werner S. Pluhar (Indianapolis, Ind.: Hackett, 1996), A32/B48.
17. Quoted in James Knowlson, *Damned to Fame: The Life of Samuel Beckett* (London: Bloomsbury, 1996), 352.
18. Michael Nyman, *Experimental Music: Cage and Beyond,* 2nd ed. (Cambridge: Cambridge University Press, 1999), 29.
19. Ibid., 9.
20. Ibid., 22n.
21. Cited in John Cage, *Silence: Lectures and Writings* (Middletown, Conn.: Wesleyan University Press, 1961), 67.
22. "The fact is that every writer creates his own precursors. His work modifies our conception of the past, as it will modify the future." Jorge Luis Borges, *Other Inquisitions, 1937–1952* (Austin: University of Texas Press, 1964), 108. "I rather think that the past doesn't go A B C, that is to say from Ives to someone younger than Ives to people still younger, but rather that we live in a field situation in which by our actions, by what we

do, we are able to see what other people do in a different light than we do without our having done anything. What I mean to say is that the music we are writing now influences the way in which we hear and appreciate the music of Ives more than that the music of Ives influences us to do what we do." John Cage, cited in Michael Nyman, *Experimental Music: Cage and Beyond*, 2nd ed. (Cambridge: Cambridge University Press, 1999), 31.

23. Karlheinz Stockhausen, *Towards a Cosmic Music*, texts selected and translated by Tim Nevill (Shaftesbury: Element, 1989), 16–17.
24. Cited in *Sound and Music in Film and Visual Media: A Critical Overview*, ed. Graeme Harper (New York: Bloomsbury, 2009), 122–23.
25. Cited in Douglas Kahn, *Earth Sound Earth Signal: Energies and Earth Magnitude in the Arts* (Berkeley: University of California Press, 2013), 101.
26. Ibid.
27. Eduardo Reck Miranda, "Brain-Computer Music Interface for Composition and Performance," *International Journal on Disability and Human Development* 5 (2006): 119–25.
28. Eduardo Reck Miranda, "Brain-Computer Music Interfacing: Interdisciplinary Research at the Crossroads of Music, Science and Biomedical Engineering," in *Guide to Brain-Computer Music Interfacing*, ed. Eduardo Reck Miranda and Julien Castet, (New York: Springer, 2014).
29. Mick Grierson, "Composing with Brainwaves: Minimal Trial P300 Recognition as an Indication of Subjective Preference for the Control of a Musical Instrument," *Proceedings of the ICMC* (2008); Yee Chieh Chew and Eric Caspary, "MusEEGk: A Brain Computer Musical Interface," *CHI Extended Abstracts on Human Factors in Computing Systems* (2011): 1417–22.
30. Miranda, *Guide to Brain-Computer Music Interfacing*, 5.
31. Michio Kaku, *The Future of the Mind: The Scientific Quest to Understand, Enhance, and Empower the Mind* (New York: Doubleday, 2014), 66.
32. Cited in Chris Gourlay, "Psychic Computer Shows Your Thoughts on Screen," *The Sunday Times* (2009), 5.
33. Kant, *Critique of Judgement*, trans. Werner S. Pluhar (Indianapolis, Ind.: Hackett, 1987), 303.
34. "Camper" refers to Peter Camper, the Dutch anatomist and arts patron known primarily for his drawings of the human head designed to illustrate phrenology.
35. For an altogether different, Hegelian account of the relation between knowing and doing as it relates to art and philosophy see Christoph Menke, "Not Yet: The Philosophical Significance of Aesthetics," in *Aesthetics and Contemporary Art*.
36. Henry Allison, *Kant's Theory of Taste*, (Cambridge: Cambridge University Press, 2001), 301.
37. This is evidently Kant's "thick" conception of genius.
38. Georg Friedrich Wilhelm Hegel, *Aesthetics: Lectures on Fine Art, Volume 1*, trans. T. M. Knox (Oxford: Oxford University Press, 1975), 75.

39. Ibid.
40. Gilles Deleuze, *Two Regimes of Madness*, trans. A. Hodge and M. Taormina (New York: Semiotext(e), 2006), 312.
41. Benedetto Croce, *The Aesthetic as the Science of Expression and of the Linguistic in General,* trans. Colin Lyas (Cambridge: Cambridge University Press, 1997), 8.
42. Hegel, *Aesthetics,* 291.
43. "Conversation with Karl Popper" in Brian Magee, *Modern British Philosophy* (London: Secker and Warburg, 1971), 73.
44. Benedetto Croce, *Guide to Aesthetics,* trans. Patrick Romanell (New York: Bobbs-Merrill, 1965), 36.
45. Croce, *The Aesthetic as the Science of Expression and of the Linguistic in General,* 76.
46. LeWitt, "Paragraphs on Conceptual Art," 834.
47. Walter Benjamin, *Reflections: Essays, Aphorisms, Autobiographical Writings,* ed. Peter Demetz, trans. Edmund Jephcott (New York: Schocken), 81.
48. Jacques Derrida, *Archive Fever: A Freudian Impression,* trans. Eric Prenowitz (Chicago: University of Chicago Press, 1995), 93.
49. Cited in John Hospers, "Artistic Creativity," *The Journal of Aesthetics and Art Criticism* 43 no. 3 (1985): 243–55 at 252.
50. See, for example, Dominic Priore *Smile: The Story of Brian Wilson's Lost Masterpiece* (London: Sanctuary, 2005).
51. LeWitt, "Paragraphs on Conceptual Art," 835.
52. Dewey, *Art as Experience,* 78–79.
53. Cees Van Leeuwen, et al., "Common Unconscious Dynamics Underlie Uncommon Conscious Effects: A Case Study in the Iterative Nature of Perception and Creation" in *Modelling Consciousness Across the Disciplines,* ed. J. Scott Jordan (Lanham, Md: University Press of America, 1999), 179–218.
54. Andy Clark, *Natural-Born Cyborgs: Minds, Technologies, and the Future of Human Intelligence* (Oxford: Oxford University Press), 77.
55. N. Katherine Hayles, *How We Think: Digital Media and Contemporary Technogenesis* (Chicago, Ill.: University of Chicago Press, 2012), 92. Here she echoes Lambros Malafouris's "At the Potter's Wheel: An Argument for Material Agency" in *Material Agency: Towards a Non-Anthropocentric Approach,* ed. Carl Knappett Malafouris and Lambros Malafouris (New York: Springer, 2008).
56. Heinrich Wölfflin, *Principles of Art History: The Problem of the Development of Style in Later Art,* trans. M. D. Hottinger (New York: Dover Publications, 1950), ix.
57. Andy Clark and David J. Chalmers, "The Extended Mind," *Analysis* 58 no. 1 (1998): 7–19. See also the literature on "distributed creativity" in the social sciences, where creativity is taken to occur in the interstices between people, places, and things, e.g., Lambros Malafouris, *How Things Shape the Mind: A Theory of Material Engagement,* trans. Colin Renfrew (Cambridge, Mass.: The MIT Press, 2013); and Vlad Petre Glaveanu, *Distributed Creativity: Thinking Outside the Box of the Creative Individual (Springerbriefs in Psychology)* (New York: Springer, 2014).

58. Slavoj Žižek, *The Indivisible Remainder: On Schelling and Related Matters* (London: Verso, 2007), 231n.
59. Marshall McLuhan, *Understanding Media* (London: Routledge, 2001), 189.
60. Marshall McLuhan and Quentin Fiore, *The Medium is the Massage: An Inventory of Effects* (Berkeley, Calif.: Gingko Press, 2001), 94.
61. *Stockhausen on Music*, compiled by Robin Maconie (London: Boyars, 2000), 120.
62. Ibid., 124.
63. For instance, the digital live-art installation "Divided by Resistance," which took place at the ICA in London in 1996 and *Thought Conductor*, which first took place at the Wintermusic festival in 1997. The latter was a musical performance where the musicians responded live to a visualized manifestation of the conductor's thought processes rather than a score. See the exhibition catalog *Null Object: Gustav Metzger Thinks About Nothing*, ed. Bruce Gilchrist and Jo Joelson (London: Black Dog Publishing, 2012).
64. Derek Bailey, *Improvisation: Its Nature and Practice in Music* (New York: Da Capo Press, 1993), 101.
65. Cited in Douglas Kahn, *Earth Sound Earth Signal*, 100.
66. Clark, *Natural-Born Cyborgs*, 120.
67. From email correspondence with Mick Grierson of Goldsmiths College.

2. Intellectual Intuition and Finite Creativity

1. A recent noteworthy contribution to this discussion is Yuval Noah Harari's *Homo Deus: A Brief History of Tomorrow* (New York: Random House, 2016).
2. Sigmund Freud, *Civilization and its Discontents*, 1930 (Standard Ed. 21), 21.
3. On the question of our future immortality see, for instance, Aubrey De Gray and Michael Rae's *Ending Aging: The Rejuvenation Breakthroughs That Could Reverse Human Aging in Our Lifetime* (New York: St. Martin's Press, 2007); and Ray Kurzweil and Terry Grossman's *Transcend: Nine Steps to Living Well Forever* (New York: Rodale Books, 2009).
4. Slavoj Žižek, *The Parallax View* (Cambridge, Mass.: The MIT Press, 2006), 192–93.
5. As Fichte admits, this may be a case of "employing the same word to express two very different concepts." Johann Gottlieb Fichte, *Introductions to the Wissenschaftslehre and Other Writings (1797–1800)*, ed. and trans. David Breazeale (Indianapolis, Ind.: Hackett, 1994), 55 (471). Wherever two sets of page numbers are provided, those appearing in brackets refer to the original edition.
6. In fact Moltke S. Gram, in his article "Intellectual Intuition: The Continuity Thesis," *Journal of the History of Ideas* 42, no. 2 (1981): 287–304, has argued that Kant does not consider intellectual intuition to be a unitary concept at all, introducing it in different places to treat logically independent problems. However, going against this thesis I will be implicitly arguing that the variant strands of intellectual intuition are different, context-specific responses to the same underlying question of human finitude.
7. Immanuel Kant, *Kant's Inaugural Dissertation of 1770*, trans. W. Eckoff, reprint ed. (Indianapolis, Ind.: Kessinger, 2004), 56.
8. Immanuel Kant, *Prolegomena to Any Future Metaphysics and the Letter to Marcus Herz,*

February 1772, 2nd ed., trans. James Ellington (Indianapolis, Ind.: Hackett, 2001) 130. [My emphasis.]

9. In characterizing Kant's project in this way I am following Henry Allison's interpretation in *Kant's Transcendental Idealism: An Interpretation and Defense, Revised and Enlarged Edition* (New Haven, Conn.: Yale University Press, 2004).
10. Immanuel Kant, *Critique of Pure Reason*, trans. Werner S. Pluhar (Indianapolis, Ind.: Hackett, 1996), A19/B33.
11. This is where Fichte and Schelling depart from Kant. Both insist upon an original faculty of intellectual intuition as the underlying substrate of all experience and the founding principle of all transcendental philosophy. Intellectual intuition here is the absolutely free and unconditioned self-positing activity of the *I* which exists only in and for its own act of knowing itself.
12. *Technics and Time, 3: Cinematic Time and the Question of Malaise*, trans. Stephen Barker (Stanford, Calif.: Stanford University Press, 2011), 70. Foucault makes a very similar argument in his famous critique of man as the "empirico-transcendental doublet." See *The Order of Things* (London: Routledge, 2002), 347.
13. Martin Heidegger, *Kant and the Problem of Metaphysics*, trans. Richard Taft (Bloomington: Indiana University Press, 1997), 36. The earlier translation by James S. Churchill hit upon the clumsy but useful couplet "Ob-ject" and "E-ject" to render the crucial distinction Heiddeger draws between Gegenstand and Ent-stand. The former is that which stands *opposed* to finite intuition, the latter is that which stands *forth* from the infinite intuition. The problem with this translation, aside from its inelegance, is that "E-ject" implies a moving outside from inside—the free production of an object [Gegenstand].
14. As Sartre notes, "being, if it is suddenly placed outside the subjective by the fulguration of which Leibniz speaks, can only affirm itself as distinct from and opposed to its creator; otherwise it dissolves in him. The theory of perpetual creation, by removing from being what the Germans call *Selbständigkeit*, makes it disappear in the divine subjectivity. If being exists as over against God, it is its own support; it does not preserve the least trace of divine creation. In a word, even if it had been created, being-in-itself would be *inexplicable* in terms of creation; for it assumes its being beyond the creation." *Being and Nothingness*, trans. Hazel E. Barnes, (London: Routledge, 2003), 20.
15. Alva Noë, *Varieties of Presence* (Cambridge, Mass.: Harvard University Press), 42.
16. Heidegger, *Kant and the Problem of Metaphysics*, 22.
17. Immanuel Kant, *Critique of Judgement*, trans. Werner S. Pluhar (Indianapolis, Ind.: Hackett, 1987), 401–402.
18. Heidegger, "Kant's Thesis About Being," trans. Ted E. Klein Jr. and William E. Pohl, in *Pathmarks* (Cambridge: Cambridge University Press, 1998), 358.
19. Immanuel Kant, *Anthropology from a Pragmatic Point of View*, trans. Victor Lyle Dowdell, revised and updated by Hans H. Rudnick (Carbondale: Southern Illinois University Press, 1996), 167.
20. Heidegger, *Kant and the Problem of Metaphysics*, 92.
21. Kant, *Anthropology from a Pragmatic Point of View*, 167–68.

22. Cited in Martin Weatherston, *Heidegger's Interpretation of Kant* (New York: Palgrave Macmillan, 2002), 175.
23. For an account of the origins of the resuscitation of the idea of intellectual intuition in Salomon Maimon's critique of Kant and his profound influence on the course of German Idealism, see Frederick C. Beiser, *The Fate of Reason: German Philosophy from Kant to Fichte* (Cambridge, Mass.: Harvard University Press, 1987).
24. Martin Heidegger, *Being and Time: A Translation of Sein und Zeit,* trans. Joan Stambaugh (Albany, N.Y.: SUNY Press), 262 (285).
25. Ibid., 136 (145).
26. Heidegger, "On the Essence of Ground," trans. William McNeill, in *Pathmarks*, 128.
27. Ibid., 128–29.
28. Ibid.
29. Heidegger, *The Basic Problems of Phenomenology*, trans. Albert Hofstadter (Bloomington, Ind.: Indiana University Press, 1988), 88.
30. Sartre, *Being and Nothingness*, 351.
31. Ibid.
32. Ibid., 504.
33. Thomas Metzinger, *Being No One: The Self-Model Theory of Subjectivity* (Cambridge, Mass.: The MIT Press, 2003), 329.
34. Ibid.
35. Thomas Bernhard, *Correction*, trans. Sophie Wilkins, (London: Vintage, 2003), 148.
36. Hubert L. Dreyfus, *Being-in-the-World: A Commentary on Heidegger's* Being and Time *Division 1* (Cambridge, Mass.: The MIT Press, 1991), 85.
37. The irony, which has been repeatedly pointed out ever since the *Critique's* first publication, is that Kant's solution to the problem which originates in the application of the category of causality beyond the bounds of possible experience itself commits that same error when it covertly relies on a concept of cause to account for how the thing in itself relates to appearance.
38. Jacques Derrida, "Force of Law," trans. Mary Quaintance, in *Acts of Religion*, ed. Gil Anidjar, (London: Routledge, 2002), 233.
39. See Jacques Derrida, "Before the Law," trans. Avital Ronell and Christine Roulston, in *Acts of Literature*, ed. Derek Atridge, (London: Routledge, 1992), in which Derrida undertakes a reading of Kafka's famous parable of the same title: "For the law is prohibition/prohibited [*inderdit*]. Noun and attribute. Such would be the terrifying double-bind of its own taking-place. It is prohibition: this does not mean that it prohibits, but that it is itself prohibited, a prohibited place. It forbids itself and contradicts itself by placing the man in its own contradiction: one cannot reach the law, and in order to have a *rapport* of respect with it, *one must not* have a rapport with the law, *one must interrupt the relation*" (203–4).
40. Immanuel Kant, *Critique of Practical Reason,* trans. Werner S. Pluhar (Indianapolis, Ind.: Hackett, 2002), 82.
41. Ibid., 83.

42. Slavoj Žižek, *Tarrying with the Negative: Kant, Hegel and the Critique of Ideology* (Durham, N.C.: Duke University Press, 1993), 14.
43. Ibid.
44. Ibid.
45. Ibid., 12.
46. Ibid., 17.
47. Kant, *Critique of Practical Reason,* 146–47.
48. Žižek, *The Parallax View*, 23.
49. On this point see Derrida's treatment of the aporia of responsibility in *The Gift of Death*, trans. David Wills (Chicago, Ill.: University of Chicago Press, 1996) and the aporetic structure of the decision in *The Politics of Friendship* trans. George Collins (London: Verso, 2005).
50. Heidegger, "The Origin of the Work of Art," trans. Albert Hofstadter, in *Basic Writings*, ed. David Farrell Krell (London: HarperCollins, 2008), 178.
51. Quentin Meillassoux, *After Finitude: An Essay on the Necessity of Contingency*, trans. Ray Brassier (London: Continuum, 2008), 5.
52. The relationship between correlationism and science is not as straightforwardly antagonistic as Meillassoux presents it. As Francisco Varela, Evan Thompson and Eleanor Rosch show in their book *The Embodied Mind* (Cambridge, Mass.: The MIT Press, 1991), rather than standing in the way between science and its object, the "structural coupling" (Varela's term) of mind and world itself forms the object of mature cognitive science and artificial intelligence research programs: "If we are forced to admit that cognition cannot be properly understood without common sense, and that common sense is none other than our bodily and social history, then the inevitable conclusion is that knower and known, mind and world, stand in relation to each other through mutual specification or dependent coorigination" (150).
53. Heidegger, *Basic Problems of Phenomenology*, 165.
54. Ibid.
55. Heidegger, *Being and Time*, 196 (212).
56. Heidegger, *Basic Problems of Phenomenology*, 170.
57. As Peter Hallward has argued, Meillassoux's critique of correlationism "seems to depend on an equivocation regarding the relation of thinking and being, of epistemology and ontology." Peter Hallward, "Anything is Possible: A Reading of Quentin Meillassoux's *After Finitude*," in *The Speculative Turn: Continental Materialism and Realism*, eds. Levi Bryant, Nick Srnicek, and Graham Harman, (Melbourne: re.press, 2011), 137.
58. Meillassoux does not shy away from the dizzying and counter-intuitive consequences of this commitment to absolute unreason and has an intricate and compelling argument to respond to the inevitable objection: namely, why, given the nonnecessity of physical laws, do we experience such remarkable stability? If nothing holds them in place, why do the laws of nature not constantly change? Such an inference, Meillassoux argues, rests on "probabilistic reasoning" that confuses contingency with chance. His refutation of this inference rests on the application of the category of the transfinite, ap-

propriated from Cantorian set-theory. See the chapter "Hume's Problem" in *After Finitude*, 82–111.

59. That Meillassoux is "*assertoric* where he should be *argumentative*" is a criticism made by Justin Clemens in his essay "Vomit Apocalypse; Or, Quentin Meillassoux's *After Finitude*," *Parrhesia* 18 (2013): 57–67.
60. Ray Brassier, "The Enigma of Realism: On Quentin Meillassoux's *After Finitude*," *Collapse* 2 (2007): 46.
61. Meillassoux, "Speculative Realism," *Collapse* 3 (2007): 433.
62. Ibid., 433–34.
63. Alberto Toscano, "Against Speculation, or, A Critique of the Critique of Critique: A Remark on Quentin Meillassoux's *After Finitude* (After Colletti)" in *The Speculative Turn*, 91.

3. *Unus Mundus*

1. Sigmund Freud, "Civilization and its Discontents," in *The Standard Edition of the Complete Psychological Works of Sigmund Freud*, ed. and trans. James Strachey (London: Hogarth Press, 1961), 21: 64.
2. Ibid., 67.
3. Ibid.
4. Ibid., 68.
5. Cf. "Beyond the Pleasure Principle," in *The Standard Edition of the Complete Psychological Works of Sigmund Freud*, ed. and trans. James Strachey (London: Hogarth Press, 1955), 18.
6. Sigmund Freud, "The Interpretation of Dreams," in *The Standard Edition of the Complete Psychological Works of Sigmund Freud*, ed. and trans. James Strachey (London: Hogarth Press, 1958), 5: 565.
7. Ibid., 566
8. Ibid.
9. Ibid.
10. Ibid. It is clear from the example of hunger that this model can only be an ideal one presented as a useful fiction. The system is always already dependent on that which is outside of it.
11. Fabio Babiloni, et al., "On the Use of Brain–Computer Interfaces Outside Scientific Laboratories: Toward an Application in Domotic Environments," *International Review of Neurobiology* 86 (2009): 133–46.
12. Fabio Aloise, et al., "Controlling Domotic Appliances via a 'Dynamical' P300-based Brain Computer Interface" in *Assistive Technology from Adapted Equipment to Inclusive Environments*, eds. P.L. Emiliani, et al. (Amsterdam: IOS Press, 2009).
13. Sigmund Freud, "Formulations on the Two Principles of Mental Functioning," in *The Standard Edition of the Complete Psychological Works of Sigmund Freud*, ed. and trans. James Strachey (London: Hogarth Press, 1958), 12: 221.
14. "The Dissection of the Psychical Personality." in *The Standard Edition of the Complete*

Psychological Works of Sigmund Freud, ed. and trans. James Strachey (London: Hogarth Press, 1964), 22: 76.

15. Freud, "Formulations on the Two Principles of Mental Functioning," 225.
16. It is well known that Freud never ceased to modify and revise the concept of the pleasure principle. Up until very late in his life it was held to be coextensive with the ancient aim of minimizing stimulation, but as he later admits, "there are pleasurable tensions and unpleasurable relaxations of tension." ("The Economic Problem of Masochism," in *The Standard Edition of the Complete Psychological Works of Sigmund Freud*, ed. and trans. James Strachey (London: Hogarth Press, 1961), 19: 160.) With the introduction of the death drive and its opposition to the libidinal (life) drives, the problem was resolved: the principle of constancy is tied to the death drive—the drive toward an inanimate state—while the pleasure principle represents a modification—from quantity to quality—of the principle of constancy due to the demands of the libidinal drives. The reality principle supervenes onto these two principles, thus completing the picture. (Ibid.)
17. Jacques Derrida, "To Speculate—On 'Freud,'" *The Post Card: From Socrates to Freud and Beyond*, trans. Alan Bass (Chicago, Ill.: University of Chicago Press, 1987), 284.
18. Freud, "Civilization and its Discontents," 76.
19. Ibid., 83.
20. Freud, "Beyond the Pleasure Principle," 38.
21. See Nick Bostrom's "Letter from Utopia," *Studies in Ethics, Law, and Technology* 2, no. 1 (2008): 1–7, where we read in flowery, overwrought prose about the perpetual state of bliss experienced by our technologically and chemically enhanced distant descendants. We are asked to "imagine a moment of bliss" and then the feeling of disillusioned despair when it fades, only to be told that such joyful moments are "not close to what I [our post-human cousin writing from 'utopia'] have now—a beckoning scintilla at most." Yet without its contrast with the feelings which preceded and followed it, what *is* that moment of bliss? How could it be maintained indefinitely without becoming everyday mundanity? As Wertheimer, the title character in Bernhard's *The Loser* says, "unhappiness is the precondition for the fact that we can be happy, only through the detour of unhappiness can we be happy." Thomas Bernhard, *The Loser*, trans. Jack Dawson (New York: Vintage, 1991), 64.
22. Paul Virilio, *Open Sky*, trans. Julie Rose (London: Verso: 1997), 20.
23. Slavoj Žižek, *The Parallax View* (Cambridge, Mass.: The MIT Press, 2009), 192.
24. Bernard Stiegler, *What Makes Life Worth Living: On Pharmacology*, trans. Daniel Ross (Cambridge: Polity Press, 2013), 30.
25. Freud, "Negation," in *The Standard Edition of the Complete Psychological Works of Sigmund Freud*, ed. and trans. James Strachey (London: Hogarth Press, 1961), 19: 237.
26. "The Uncanny," in *The Standard Edition of the Complete Psychological Works of Sigmund Freud*, ed. and trans. James Strachey (London: Hogarth Press, 1955), 17: 239.
27. Ibid., 240.
28. Ibid.

29. "A Metapsychological Supplement to the Theory of Dreams," in *The Standard Edition of the Complete Psychological Works of Sigmund Freud*, ed. and trans. James Strachey (London: Hogarth Press, 1957), 14: 231.
30. Ibid., 223.
31. Ibid., 233.
32. "Fragment of an Analysis of a Case of Hysteria," in *The Standard Edition of the Complete Psychological Works of Sigmund Freud*, ed. and trans. James Strachey (London: Hogarth Press, 1953), 7: 110.
33. "The Paths to the Formation of Symptoms," in *The Standard Edition of the Complete Psychological Works of Sigmund Freud*, ed. and trans. James Strachey (London: Hogarth Press, 1963), 16: 358.
34. Tarkovsky's psychoanalytically-informed understanding of what would constitute an "innermost wish" differs crucially from the Strugatsky brothers' original story. There it is described as, " 'the kind that, if they don't come true, you'd be ready to jump off a bridge!' ", while in the film it is the coming true itself that would be unbearable. Arkady and Boris Strugatsky, *Roadside Picnic*, trans. Olena Bormashenko (London: Gollancz), 165.
35. "The Paths to the Formation of Symptoms," 372.
36. "A Note Upon the 'Mystic Writing-Pad,' " in *The Standard Edition of the Complete Psychological Works of Sigmund Freud*, ed. and trans. James Strachey (London: Hogarth Press, 1961), 19: 228.
37. Mark B. N. Hansen, *Feed-Forward: On the Future of Twenty-First Century Media* (Chicago, Ill.: University of Chicago Press, 2015), 54.
38. Amy Kruse, *Operational Neuroscience: Intelligence Community Forum* (05.11.08) http://www.dtic.mil/ndia/2008intell/kruse.pdf.
39. Hansen, *Feed-Forward*, 55.
40. Ibid., 138.
41. Ibid., 58.
42. Martin Heidegger, *The Question Concerning Technology and Other Essays*, trans. William Lovitt (London: Harper & Row, 1977), 20
43. Elizabeth A. Phelps, et al., "Performance on Indirect Measures of Race Evaluation Predicts Amygdala Activation," *Journal of Cognitive Neuroscience* 12, no. 5 (2000): 729–38; David M. Amodio, Eddie Harmon-Jones, and Patricia G. Devine, "Individual Differences in the Activation and Control of Affective Race Bias as Assessed by Startle Eyeblink Responses and Self-report," *Journal of Personality and Social Psychology* 84, no. 4 (2003): 738–53.
44. Anthony Greenwald of the University of Washington is the figure credited with pioneering this method. See Anthony G. Greenwald, Debbie E. McGhee, and Jordan L. K. Schwartz, "Measuring Individual Differences in Implicit Cognition: The Implicit Association Test." *Journal of Personality and Social Psychology* 74, no. 6 (1998): 1464–80. IATs can be easily found online and are used to measure implicit attitudes, identities, and stereotypes. The racial stereotyping variants of the test have often captured the

attention of the mainstream media, being taken as evidence that "we are all racists at heart," as the title of a *Wall Street Journal* article of December 2005 declares. The danger, as demonstrated by this same article, is that such studies are used ideologically to naturalize prejudice and discrimination, in exactly the same way that biological findings are distorted to naturalize inequality between the sexes.

45. As an extreme example of this discrepancy see the fascinating case of James Fallon, the neuroscientist who discovered by accident, from a scan of his own brain, that he had the genetic makeup of a psychopath. The shock of this discovery was compounded when he learned that friends and relatives indeed considered him to have psychopathic traits. James Fallon, *The Psychopath Inside: A Neuroscientist's Personal Journey into the Dark Side of the Brain* (New York: Penguin, 2014).
46. Žižek, *The Parallax View*, 170.
47. Slavoj Žižek, *The Plague of Fantasies* (London: Verso, 2008), 156.
48. However, transforming this very "understanding of ourselves" in line with current knowledge about the brain is precisely the project of eliminative materialism, as espoused most notably by Patricia and Paul Churchland. They argue that if we cannot incorporate such neuroscientific knowledge into the existing categories that we use to describe conscious experience it is because these latter are *false*. See Paul Churchland's *Matter and Consciousness* revised ed. (Cambridge, Mass: The MIT Press, 1988) for an accessible introduction to this position.
49. As Paul Churchland has said in an interview, with regard to the just-mentioned categories he seeks to displace (namely propositional attitudes explaining human behavior through recourse to belief, desire, free will, etc.), "Now of course, that is how I think of myself; in terms of propositional attitudes! I am as much in the grip of that conception as anyone else." "Demons Get Out!: Interview with Paul Churchland," *Collapse* 2 (2007), 209.
50. Ibid., 157.
51. Ibid., 158.
52. Žižek, *The Parallax View*, 172. Adrian Johnston has written fascinatingly on Freud and Lacan's conflicting attitudes to the possibility of unconscious affects and the possibility for feelings to be "misfelt": that is, for us to be misled as to the true nature of our most intimate, apparently immediate affects. See his contribution to the jointly authored volume with Catherine Malabou, *Self and Emotional Life: Philosophy, Psychoanalysis, and Neuroscience* (New York: Columbia University Press, 2013).
53. Slavoj Žižek, *The Ticklish Subject: The Absent Centre of Political Ontology* (London: Verso, 1999), 301.
54. Ibid.
55. Ibid., 302.
56. Ibid.
57. Sigmund Freud, "Repression," in *The Standard Edition of the Complete Psychological Works of Sigmund Freud*, ed. and trans. James Strachey (London: Hogarth Press, 1957), 14: 147.

58. Ibid., 148.
59. Ibid.
60. Ibid., 149.
61. Sigmund Freud, "The Unconscious," in *The Standard Edition of the Complete Psychological Works of Sigmund Freud*, ed. and trans. James Strachey (London: Hogarth Press, 1957), 14: 201.
62. Michael S. Gazzaniga, *Who's in Charge? Free Will and the Science of the Brain* (New York: Ecco, 2012), 83. Obviously this experiment can only be performed on split-brain patients as, in a typical patient, it is not possible to present information only to one hemisphere of the brain.
63. *The Pervert's Guide to Cinema*, directed by Sophie Fiennes (2006), episode 1.
64. Freud "Beyond the Pleasure Principle," 28.
65. Sigmund Freud, *The Psychopathology of Everyday Life*, trans. Anthea Bell (London: Penguin, 2002), 265n.
66. Freud, "The Interpretation of Dreams," 578.
67. Freud, "Beyond the Pleasure Principle," 28.
68. Freud, "The Ego and the Id," in *The Standard Edition of the Complete Psychological Works of Sigmund Freud*, ed. and trans. James Strachey (London: The Hogarth Press, 1961), 19: 55.
69. Freud, "The Dissection of the Psychical Personality," 74.
70. See the work of Alain Brunet at McGill University cited in Emily Singer, "Manipulating Memory," *MIT Technology Review*: https://www.technologyreview.com.
71. Once again a productive connection could be drawn with the "times" of experience in Mark Hansen's *Feed-Forward*—the slow time of consciousness, as opposed to the microtemporal scales of subperceptual experience. The two temporal dimensions are utterly incommensurable and can only communicate via the translation mechanism Hansen names "feed-forward." In this sense, psychoanalysis itself could be considered a process of feeding-forward.
72. Jacques Derrida, *Writing and Difference*, trans. Alan Bass (London: Routledge, 2001), 270.
73. "The Paths to the Formation of Symptoms," 371. "Moses and Monotheism" is the text in which this thesis finds its fullest expression.
74. Freud's late endorsement of Lamarckian inheritance has not been well received, either by his followers or by his critics, who see it as flying in the face of science. However, recent developments in the science of epigenetics lead us to reevaluate the biological grounds on which Freud's (and Jung's) findings can be dismissed as irrational. See my "Bernard Stiegler on Transgenerational Memory and the Dual Origin of the Human," *Theory, Culture and Society* 33, no. 3 (May 2016): 151–73 for a fuller development of this argument.
75. See "From the History of an Infantile Neurosis," 97.
76. Ibid., 65.
77. Carl Gustav Jung, "The Psychology of the Unconscious," in *The Collected Works of C. G.*

Jung, 2nd ed., ed. and trans. R. F. C. Hull (London: Routledge & Kegan Paul, 1966), 7: 64–65.

78. Ibid., 161.
79. Ibid., 175.
80. Ibid.
81. As Derrida wrote, "A certain alterity—to which Freud gives the metaphysical name of the unconscious—is definitively exempt from every process of presentation by means of which we would call upon it to show itself in person." *Margins of Philosophy*, trans. Alan Bass (Chicago, Ill.: University of Chicago Press, 1982), 20.
82. Jung, "The Psychology of the Unconscious," 183.
83. Ibid., 175.
84. Carl Gustav Jung, "Archetypes of the Collective Unconscious," in *The Collected Works of C. G. Jung*, 2nd ed., ed. and trans. R. F. C. Hull (London: Routledge & Kegan Paul, 1968), 9.1: 5.
85. Jung was heavily influenced by Kant, particularly by the famous pre-Critical work *Dreams of a Spirit-Seer*, in which Kant allows himself, often in an ironic, uncharacteristically playful fashion, to speculate upon that which lies beyond the bounds of human experience and knowledge through a satirical reading of the work of Swedish mystic Emmanuel Swedenborg.
86. "On the Nature of the Psyche," in *The Collected Works of C. G. Jung*, 2nd ed., ed. and trans. R. F. C. Hull (London: Routledge & Kegan Paul, 1969), 8: 205.
87. Ibid., 96–97.
88. "The Psychology of the Unconscious," 93.
89. Ibid., 107.
90. "Synchronicity: An Acausal Connecting Principle," in *The Collected Works of C. G. Jung*, 2nd ed., ed. and trans. R. F. C. Hull (London: Routledge & Kegan Paul, 1969), 8: 438.
91. While Jung makes frequent, approving references to philosophical forebears such as Schopenhauer, Leibniz, Kant, Lao-Tzu, and others, one name entirely absent from the *Synchronicity* book is Spinoza. Nevertheless, Jung's psychoid ontology carries clear echoes of Spinoza's dual aspect monism. For example, this passage from the *Ethics*: "the mind and the body are the same thing, which is conceived now under the attribute of thought, now under the attribute of extension. The result is that the order, *or* connection, of things is one, whether Nature is conceived under this attribute or that." Benedict de Spinoza, *Ethics*, trans. Edwin Curley (London: Penguin Books, 1996), 71 (II/141).
92. As a caveat it should be noted that panpsychism is a far from homogeneous position and the charge of dualism would not necessarily apply to all of its variants. See in particular Steven Shaviro, *The Universe of Things: On Speculative Realism* (Minneapolis: University of Minnesota Press, 2014).
93. Quentin Meillassoux, *After Finitude: An Essay on the Necessity of Contingency*, trans. Ray Brassier (London: Continuum, 2008), 37.
94. Paul Bishop, *Synchronicity and Intellectual Intuition in Kant, Swedenborg, and Jung*

(Lewiston, N.Y.: The Edwin Mellen Press, 2000). Bishop's primary thesis is that Jung's work on synchronicity is based on a misprision, or creative misunderstanding of Kant.

95. Cited in John Hospers, "Artistic Creativity," *The Journal of Aesthetics and Art Criticism* 43, no. 3 (Spring, 1985): 252.
96. Jean-François Lyotard, *The Inhuman: Reflections on Time*, trans. Geoffrey Bennington and Rachel Bowlby (Cambridge: Polity Press), 18.
97. Friedrich Nietzsche, *Ecco Homo*, trans. R. J. Hollingdale (London: Penguin, 1988), 102.
98. For a fuller development of this logic, via a reading of Derrida and Kant, I refer the reader to my article "Genius Is What Happens: Derrida and Kant on Genius, Rule-Following and the Event," *British Journal of Aesthetics* 54, no. 3 (2014): 323–37.
99. Carles Grau, et al., "Conscious Brain-to-Brain Communication in Humans Using Non-Invasive Technologies," *PLOS ONE* 9, no. 8 (2014).
100. David Roden, *Posthuman Life: Philosophy at the Edge of the Human* (London: Routledge, 2015), 3.

4. Techno-Telepathy and the Otherness of the Other

1. It has, however, become a significant theme in recent literary theory and cultural studies discourses, due to the important work of Nicholas Royle, both in his capacity as English translator of Derrida's essay "Telepathy" (to which we will return), and also through his own writings, in particular *Telepathy and Literature: Essays on the Reading Mind* (Oxford: Blackwell, 1991); his specially edited issue of *Oxford Literary Review* 30, no. 2 (2008), titled "Telepathies"; the essay "The Remains of Psychoanalysis (i): Telepathy", in *After Derrida* (Manchester: Manchester University Press, 1995); and the chapter on "The Telepathy Effect" in *The Uncanny: An Introduction* (Manchester: Manchester University Press, 2002). See also J. Hillis Miller's entertaining short book *The Medium is the Maker: Browning, Freud, Derrida and the New Telepathic Ecotechnologies* (Brighton: Sussex Academic Press, 2009); and Martin McQuillan's essay "Tele-Techno-Theology" in his *Deconstruction After 9/11* (London: Routledge, 2009), 47–64. On the cultural history of telepathy, its origins and "extraordinary persistence," see Roger Luckhurst, *The Invention of Telepathy: 1870–1901* (Oxford: Oxford University Press, 2002).
2. Freeman Dyson, "'Radiotelepathy,' The Direct Communication of Feelings and Thought from Brain to Brain," *Edge* (2009): http://www.edge.org.
3. In the existing scientific literature the difference between "brain-to-brain" and "mind-to-mind" communication is merely that between conscious and unconscious transmission.
4. Rajesh P. N. Rao et al., "A Direct Brain-to-Brain Interface in Humans," *PLOS ONE* 9, no. 11 (2014).
5. Seung-Schik Yoo et al., "Non-Invasive Brain-to-Brain Interface (BBI): Establishing Functional Links between Two Brains," *PLOS ONE* 8, no. 4 (2013).
6. John B. Trimper, Paul Root Wolpe, and Karen S. Rommelfanger, "When 'I' Becomes 'We': Ethical Implications of Emerging Brain-to-Brain Interfacing Technologies," *Frontiers in Neuroengineering* 7, no. 4 (2014).

7. Andrea Stocco et al., "Playing 20 Questions with the Mind: Collaborative Problem Solving by Humans Using a Brain-to-Brain Interface," *PLOS ONE* 10, no. 9 (2015).
8. Ibid.
9. Carles Grau et al., "Conscious Brain-to-Brain Communication in Humans Using Non-Invasive Technologies," *PLOS ONE* 9, no. 8 (2014).
10. Ibid.
11. Miguel Pais-Vieira et al., "A Brain-to-Brain Interface for Real-Time Sharing of Sensorimotor Information," *Scientific Reports* 3 (2013).
12. Miguel Pais-Vieira et al., "Building an Organic Computing Device with Multiple Interconnected Brains," *Scientific Reports* 5 (2015).
13. Arjun Ramakrishnan et al., "Computing Arm Movements with a Monkey Brainet," *Scientific Reports* 5 (2015).
14. Pais-Vieira et al., "Building an Organic Computing Device with Multiple Interconnected Brains."
15. Jessica Hamzelou interview with Andrea Stocco, "Let's Talk, Brain to Brain," *New Scientist* 3063 (5 March 2016).
16. As we will discuss below, however, fully realized thought communication represents the very limit of telecommunication.
17. Grau et al., "Conscious Brain-to-Brain Communication in Humans Using Non-Invasive Technologies."
18. Slavoj Žižek, *The Parallax View* (Cambridge, Mass.: The MIT Press, 2009), 178.
19. Jacques Derrida, *Psyche: Inventions of the Other Volume 1* (Stanford, Calif.: Stanford University Press), 226.
20. Immanuel Kant, *Critique of Practical Reason*, trans. Werner S. Pluhar (Indianapolis, Ind.: Hackett, 2002), 99.
21. Ibid.
22. Emmanuel Levinas, *Totality and Infinity: An Essay on Exteriority*, trans. Alphonso Lingis (Pittsburgh, Pa.: Duquesne University Press, 1969), 73.
23. Ibid.
24. I owe this formulation to a conversation with Alex Düttmann.
25. "Murder exercises a power over what escapes power. It is still a power, for the face expresses itself in the sensible, but already impotency, because the face rends the sensible." Levinas, *Totality and Infinity*, 98.
26. Søren Kierkegaard, *Fear and Trembling*, trans. Alastair Hannay (London: Penguin, 1986), 89.
27. Ibid., 109.
28. Jacques Derrida, *The Gift of Death*, trans. David Wills (Chicago, Ill.: University of Chicago Press, 1996), 60.
29. Jacques Derrida, *Speech and Phenomena and Other Essays on Husserl's Theory of Signs*, trans. David B. Allison (Evanston, Ill.: Northwestern University Press), 31.
30. Jacques Derrida, *Margins of Philosophy*, trans. Alan Bass (Chicago, Ill.: University of Chicago Press), 16.

31. Ludwig Wittgenstein, *Philosophical Investigations*, trans. GEM Anscombe (Oxford: Blackwell, 1972), 108.
32. Ibid., 109.
33. David Roden, *Posthuman Life: Philosophy at the Edge of the Human* (London: Routledge, 2015), 4.
34. Ibid.
35. Ibid.
36. As well as anticipating canonical discussions in twentieth century analytic philosophy concerning the relation between thought and language (or "thought and talk" as in the title of Donald Davidson's famous essay), and the possibility of a private language, it also resonates with recent work on the social construction of consciousness—see T. R. Burns and E. Engdahl, "The Social Construction of Consciousness: Part 2: Individual Selves, Self-awareness, and Reflectivity," *Journal of Consciousness Studies* 5, no. 2 (1998): 166–84.
37. Friedrich Nietzsche, *The Gay Science with a Prelude in Rhymes and an Appendix of Songs*, trans. Walter Kaufmann. (New York: Vintage, 1974), 298.
38. Sigmund Freud, "The Ego and the Id," in *The Standard Edition of the Complete Psychological Works of Sigmund Freud*, ed. and trans. James Strachey (London: The Hogarth Press, 1961), 19: 23.
39. Sigmund Freud, "The Unconscious," in *The Standard Edition of the Complete Psychological Works of Sigmund Freud*, ed. and trans. James Strachey (London: Hogarth Press, 1957), 14: 194.
40. Sigmund Freud, "Dreams and Occultism," in *The Standard Edition of the Complete Psychological Works of Sigmund Freud*, ed. and trans. James Strachey (London: The Hogarth Press, 1964), 22: 55. As support for this hypothesis he cites the seemingly inexplicable common purpose demonstrated by insect communities, much in the way that Jung refers to the communicative dances of the bees to demonstrate the possibility of transcerebral thought and communication. Freud's other writings on telepathy include "Psycho-Analysis and Telepathy" and "Dreams and Telepathy" in *The Standard Edition of the Complete Psychological Works of Sigmund Freud*, ed. and trans. James Strachey (London: The Hogarth Press, 1955), 18: 177–93; 197–220.
41. Wittgenstein, *Philosophical Investigations*, 153.
42. Husserl, *Cartesian Meditations*, 37. Similarly, in the *Ideas I*, Husserl writes that "*no concrete mental process* can be accepted as *a self-sufficient one in the full sense*," which means that no "content" of phenomenological experience can be taken on its own, for each moment of consciousness is what it is only within its contextual "stream." Furthermore, "two *streams of* mental processes (spheres of consciousness for two pure Egos) *of an identically essential content are inconceivable*, as well as that no *completely determined* mental process of the one stream can belong to the other . . . only mental processes of an identical inner characteristic can be common to them (although not common as individually identical), but not two mental processes which, in addition, have a 'halo' absolutely alike." Edmund Husserl, *Ideas Pertaining to a Pure Phenomenology*

and to a Phenomenological Philosophy, First Book, trans. F. Kersten (The Hague: Martinus Nijhoff, 1982), 198–99.

43. Wittgenstein, *Philosophical Investigations*, 153.
44. Edmund Husserl, *Cartesian Meditations: An Introduction to Phenomenology*, trans. Dorion Cairns (The Hague: Martinus Nijhoff, 1977), 104.
45. This "background" of associations and memories is closely related to what in cognitive science is known as "the frame problem," that is, the immense difficulty in replicating in a computational system the ease and speed with which we find our bearings in a situation and process or make sense of any event with which we are faced, however unexpected. Since we have gathered such an immense store of knowledge over the course of our lives, and everything we have learned is potentially relevant in any situation, this knowledge must be immediately accessible at all times so that it can be called upon as soon as it is required. Replicating this ability in a strong AI program would impose huge demands on the structuring and processing of information.
46. "What someone once said of Homer—that to understand him well means to admire him—is also true for the art works of the ancients, especially the Greeks. One must become as familiar with them as with a friend in order to find their statue of Laocoon just as inimitable as Homer. In such close acquaintance one learns to judge as Nicomachus judges Zeuxis' Helena: 'Behold her with my eyes,' he said to an ignorant person who found fault with this work of art, 'and she will appear a goddess to you.'" Johann Joachim Winckelmann, *Reflections on the Imitation of Greek Works in Painting and Sculpture*, trans. Elfriede Heyer and Roger C. Norton (Chicago, Ill.: Open Court Publishing, 1987), 5.
47. In §302 of the *Philosophical Investigations*, Wittgenstein writes: "If one has to imagine someone else's pain on the model of one's own, this is none too easy a thing to do: for I have to imagine pain which I *do not feel* on the model of the pain which I *do feel*. That is, what I have to do is not simply to make a transition in imagination from one place of pain to another. As, from pain in the hand to pain in the arm. For I am not to imagine that I feel pain in some region of his body. (Which would also be possible) (101)." This is an exceedingly tricky passage to fathom, seeming at once completely straightforward and bafflingly obtuse. For I can and do all the time imagine pain I do not feel "on the model of" pain I do feel, as when I wince at the sight of a nasty injury. However, following Søren Overgaard's reading in his excellent book *Wittgenstein and Other Minds: Rethinking Subjectivity and Intersubjectivity with Wittgenstein, Levinas, and Husserl* (London: Routledge 2007), we can see that what we are being asked to imagine here is how we could feel someone else's pain without making it *our own* pain; thus me feeling his pain not as my own but *as his*. Italicizing the second "I" in the sentence "For I am not to imagine that *I* feel pain in some region of his body" makes the sense, and the problem, somewhat clearer.
48. Jacques Derrida, *Writing and Difference*, trans. Alan Bass (London: Routledge, 2001), 163.

49. Husserl, *Cartesian Meditations,* 89.
50. Levinas, *Totality and Infinity,* 51.
51. It will be seen how this opposition is rendered problematic in Derrida's account of teleiopoesis in the final section.
52. More than a decade earlier, in the essay *Time and the Other,* trans. Richard A. Cohen (Pittsburgh, Pa.: Duquesne University Press, 1994), Levinas writes "experience always already signifies knowledge, light, and initiative, as well as the return of the object to the subject" (70). This is why death is likewise not an experience.
53. Derrida, *Writing and Difference,* 156.
54. Ibid.
55. Levinas, *Totality and Infinity,* 64.
56. Overgaard, *Wittgenstein and Other Minds,* 102.
57. Derrida, *Psyche: Inventions of the Other,* 239.
58. Derrida, *Speech and Phenomena,* 38.
59. Ibid.
60. Jean-Paul Sartre, *Being and Nothingness: An Essay on Phenomenological Ontology,* trans. Hazel Barnes (London & New York: Routledge, 2003), 72.
61. Derrida, *Psyche: Inventions of the Other,* 239.
62. Husserl, *Cartesian Meditations,* 121.
63. Jacques Derrida, *The Politics of Friendship,* trans. George Collins (London: Verso, 2005), 32.
64. Jacques Derrida, *H.C. for Life, That Is to Say . . . ,* trans. Laurent Milesi and Stefan Herbrechter (Stanford, Calif.: Stanford University Press, 2006), 104.
65. Derrida, *Psyche: Inventions of the Other,* 227.
66. Derrida, *H.C. for Life, That Is to Say . . . ,* 66.
67. Jorge Luis Borges, "Kafka and His Precursors" in *Other Inquisitions 1937–52,* trans. Ruth L. C. Simms (London: Souvenir Press, 1964), 108.
68. Derrida elsewhere writes of the necessity of a countersignature provided by the reader of a text, which comes to divide and complicate the simple origin of the "primary" signature, such that it cannot be said who preceded whom: "It is thus from the countersignature that a signature is properly carried off. And it is in the instant when it is thus carried off that *there is text.* You therefore no longer know which of the two partners will have signed first." Jacques Derrida, *Signéponge/Signsponge,* trans. Richard Rand (New York: Columbia University Press, 1984), 130. Also, in the roundtable discussion following his paper on Nietzsche titled "Otobiographies," Derrida says, "the signature becomes effective—performed and performing—not at the moment it apparently takes place, but only later, when ears will have managed to receive the message. In some way the signature will take place on the addressee's side, that is, on the side of him or her whose ear will be keen enough to hear my name, for example, or to understand my signature, that with which I sign." Jacques Derrida, *The Ear of the Other,* trans. Peggy Kamuf (New York: Schocken Books, 1985), 50.

69. Derrida, *Psyche: Inventions of the Other*, 228–29.
70. As Nicholas Royle, paraphrasing Derrida, notes, "[difficult] to imagine a theory of fiction, a theory of the novel, without a theory of telepathy. Starting perhaps from the hypothesis that fiction is, in some radical sense, incapable of non-telepathic representation; starting from the thought that the telepathic founds the very possibility of character, characterisation, etc.—from the 'omniscient narrator' onwards." *Telepathy and Literature*, 17. In the chapter "The 'Telepathy Effect'" in *The Uncanny*, Royle develops this notion further, suggesting that the inescapably theological motif of omniscience should be replaced by a concept of telepathy.
71. The reference to Proust has further pertinence, for one could say of *À la Recherche du Temps Perdu* that it, too, "advances backwards" and "begins at the end," owing to its often remarked-upon circular structure: at the end of the novel it will have been the book to come that is now promised, and the narrator will have been the author.
72. Derrida, *Psyche: Inventions of the Other*, 228.
73. Derrida, *The Politics of Friendship*, 32.
74. Nicholas Royle suggests that deconstruction itself can be regarded as a form of telepathy, or telepathic relation—namely, the way Derrida "inhabits" other writers and philosophers, such that it is often not clear who is speaking when. This is nowhere more true than in the essay "Telepathy" itself, where Derrida sometimes writes "as" Freud, in the first person. See Royle, *After Derrida*, 68–69.
75. Derrida, *Psyche: Inventions of the Other*, 229.
76. It is important to note that for the sake of clarity and simplicity we have been proceeding as if there were only two participants in this relationship, but this is of course an oversimplification and an idealization. Indeed Derrida stresses that it is "certainly more than two, always more than two." Ibid., 228.
77. Ibid.
78. Derrida, *H.C. for Life, That Is to Say . . .*, 66.
79. Derrida, *The Politics of Friendship*, 37.
80. Derrida, *The Gift of Death*, 41.
81. "*No one can take the other's dying away from him.* Someone can go 'to his death for an other.' However, that always means to sacrifice oneself for the other '*in a definite matter.*' Such dying for . . . can never, however, mean that the other has thus had his death in the least taken away. Every Dasein must itself actually take dying upon itself. Insofar as it 'is,' death is always essentially my own. And it indeed signifies a peculiar possibility of being in which it is absolutely a matter of the being of my own Dasein. In dying it becomes evident that death is ontologically constituted by mineness and existence." Martin Heidegger, *Being and Time*, trans. Joan Stambaugh (New York: SUNY Press), 223 (240).
82. Derrida, *The Gift of Death*, 41.
83. Transhumanists may uphold the claim that advances in biotechnology or "mind uploading" will mean that our descendants will live "forever," but even if this is granted it by no means entails that death would have been *overcome* and so would no longer

form a part of life. Any being that is not a necessary being—that is, is not God—is constantly exposed to the possibility of death, regardless of its de facto lifespan. All this would mean is that death has been deferred, perhaps indefinitely (although not infinitely). Furthermore, as Spinoza says, there is no reason to assume that a person only dies when he or she is "changed into a corpse." Benedict de Spinoza, *Ethics*, trans. Edwin Curley (London: Penguin Books, 1996), 138 (II/240). We are constantly dying and outliving ourselves, or *living-on* in the language of Derrida.

84. Heidegger, *Being and Time*, 232 (250–1).
85. Emmanuel Levinas, *Entre Nous: Thinking-of-the-Other*, trans. Michael B. Smith and Barbara Harshav (London: Continuum, 2006), 92.
86. Levinas, *Entre Nous*, 52.
87. Levinas, *Totality and Infinity*, 244.
88. Emmanuel Levinas, *Otherwise than Being*, trans. Alphonso Lingis (The Hague: Nijhoff, 1981), 57.
89. "Ethics as First Philosophy," in *The Levinas Reader*, ed. Seán Hand (Oxford: Blackwell, 1989), 84.
90. As Levinas describes this immemoriality: "Here we have, in the ethical anteriority of responsibility (for-the-other, in its priority over deliberation), a past irreducible to a hypothetical present that it once was." Levinas, *Entre Nous*, 147.
91. Levinas, *Entre Nous*, 188.
92. Ibid., 188.
93. Derrida, *The Gift of Death*, 43–44.
94. Heidegger, *Being and Time*, 232 (250).
95. Ibid., 231.
96. Jacques Derrida, *Aporias: Dying—Awaiting (One Another At) the "Limits of Truth,"* trans. Thomas Dutoit (Stanford, Calif.: Stanford University Press, 1993), 75.
97. *The Gift of Death*, 108.
98. Ibid., 109.
99. Cf., for example Jacques Derrida, *Given Time: 1. Counterfeit Money*, trans. Peggy Kamuf (Chicago, Ill.: University of Chicago Press, 1992), where we read that "*there is* gift" (10)—i.e. outside of the circle of economic exchange—but "the gift does not *exist* and does not *present* itself" (15).
100. In this there are parallels with Martin Hägglund's book *Radical Atheism: Derrida and the Time of Life* (Stanford, Calif.: Stanford University Press, 2008).
101. The inevitable upshot of correlationism, says Meillassoux, is fideism: "*by forbidding reason any claim to the absolute, the end of metaphysics has taken the form of an exacerbated return of the religious.*" *After Finitude*, trans. Ray Brassier (London: Continuum, 2008), 45.

INDEX

actuality. *See* possibility
Adorno, Theodor, 9, 124
Allison, Henry, 27, 176
almost nothing, 16–18
alpha waves. *See* brainwaves
alterity. *See* otherness
Andre, Carl, 13–14
"applied neuroscience," 8, 98
Arendt, Hannah, 1
art object, the: automatic production of, 24, 39, 43; Croce on, 32–35, 36–37; and idea, 14–15, 16, 23, 123–24; Kant on, 27, 28
autocorrect, 93–94
automation, 93–94
automatism, 75–77
avant-garde, 18–19

Bailey, Derek, 40–41
Barth, John, 16
Beckett, Samuel, 15–16, 18
Being John Malkovich, 145
Benjamin, Walter, 33, 156
Bernhard, Thomas, 70, 180n21
Birds, The, 105
body: bypassing of the, 20, 42, 46, 78, 93; remote control over another's, 128–29, 131, 135, 142, 145
Borges, Jorge Luis, 19, 154–55, 172n22
Bostrom, Nick, 180n21
brain-computer interface (BCI): and art, 10–11, 24, 26, 38–43; commercial applications of, 5, 6, 89–90, 97; domestic applications of, 90–91; history of, 3–6; and intellectual intuition, 9, 11, 43, 45–48, 60–61; for music, 11, 21–23, 42, 78; and reality testing, 91, 94–95
brain-computer music interface (BCMI). *See* brain-computer interface
Brainet, 130–31, 132, 137–38
Brainstorm, 142, 144, 162–63
brain-to-brain interface (BBI), 3, 12, 37, 125–32, 137–38, 140, 148, 157
brainwaves, 4–7, 8, 20–21, 39–40, 97. *See also* brain-computer interface; EEG
Brassier, Ray, 84, 86

Cage, John, 17–19, 172–73n22
Cartesian theater, 101
causality: as category, 62; and free will, 71–73, 77, 134–35; and *Nachträglichkeit*, 109–10; and synchronicity, 114, 115–19, 122; and teleiopoesis, 155, 157
Clark, Andy, 9, 35–36, 37, 42
cognitive imaging, 7–8, 11, 23–25, 26, 60, 78, 102, 132, 142

cognitive science, 70, 121, 178n52, 188n45
conceptual art, 13–14, 16–17, 32–33
consciousness, 70; deconstruction of, 136, 139; flux of, 49, 54, 105–6, 187n42; Nietzsche on, 137–41; privacy of, 137; science of, 100–102, 121; speed of, 98
creatio ex nihilo, 60, 61, 78–79, 118
creative process, the, 13–14, 15–16; Croce on, 29–30, 31, 33–37; Kant on, 25–28; as synchronicity, 123–25
creativity, human, 2, 46–47, 59–61, 67
Croce, Benedetto, 10, 11; on the art object, 32, 34–35, 36; and conceptual art, 32–33; and embodied cognition, 36–37; and idealism, 123–24; and improvisation, 33–35; on intuition as expression, 29–30, 36, 136; on the medium, 35–37; on nonartists, 31–32; on technical skill, 31–32; on translation, 30–31
cybernetics, 1–2

data mining, 99, 103
death: experience of, 162–63; as ground of responsibility, 158–60; as most and least proper, 163–64, 167; of the other, 160–62; overcoming of, 190–91n83
Deleuze, Gilles, 29
delimit (Derrida), 17
Derrida, Jacques: on autoaffection, 136, 139; on death, 158, 161–64; on the frame, 18; on Freud, 92, 109–10; on I and other, 145, 148–52, 156–57, 166–68; on the ideality of the archive, 33; on the law, 74; on Levinas, 148–49; on the secret self, 165–66; on teleiopoesis, 152–57, 166–67; on telepathy, 12, 133, 150–52; on thought and language, 136
desire: automatic realization of, 2, 70, 90–91, 96, 102–4, 133; Freud on, 11, 87–88, 92; operationalization of, 94, 98. *See also* wish-fulfillment
Dewan, Edmond, 3–5, 20
Dewey, John, 14, 25, 35, 37
digital media, 37–38, 96–97, 98–99
direct communication, 12, 125, 127–28, 130, 131–33, 136; of artistic ideas, 14, 20, 28, 37; of feelings, 127, 142; impossibility of, 149–50, 151–52. *See also* telepathy
divine insight, 10, 73, 76–77, 133–34, 165
Dreyfus, Hubert, 70
Dr. Mabuse, 135
Duchamp, Marcel, 14
Dyson, Freeman, 127, 132

ECoG, 6
EEG, 3–8, 21–23, 39, 90, 97, 128, 129
enhancement. *See* human enhancement
event, the: coincidence with thought, 9, 23, 40, 43, 60, 70, 91–92; of creation, 61, 71, 105, 109; of synchronicity, 118–19, 120, 124, 125, 166; of teleiopoesis, 153; of telepathy, 132
event-related potentials, 4, 8, 22
evoked potentials. *See* event-related potentials
Experimental Music, 18–19
expression. *See* creative process, the; Croce, Benedetto

Facebook, 91, 98, 127
facticity, 1, 68–69, 82–85, 122
Fichte, J. G., 2, 46, 66, 69, 175, 176
finitude, overcoming of, 1, 2, 78–79, 94–95, 110, 145, 159, 166–68; in Heidegger, 66–67, 69, 70–71; in Jung, 120–23; in Kant, 11, 46–47, 56–57, 71, 74–75, 76–78, 134; in Meillassoux, 82–84, 85–86; in Žižek, 9–10, 45–46
first-person experience, transmission of, 141–45. *See also* direct communication; telepathy
fluency, 41, 42
freedom as economy of unfreedom, 69–71, 73–78

Freud, Sigmund: on dreams, 89, 91, 95; on fantasy, 96; on individuation, 11–12, 88; on internal and external, 11–12, 87–88, 89, 91–92, 94–95; and Jung, 111, 112, 116, 117; on *Nachträglichkeit,* 109–10, 114, 155; and Nietzsche, 140–41; on the object, 87–88; on the oceanic, 87–88; on the pleasure principle, 89, 91–92, 95, 180n16; on primal fantasy, 110; on reality testing, 89, 91, 94–95, 107, 117; on repression, 91, 95, 103–5, 106, 108; on technology, 9, 45, 89, 97, 140; on the timelessness of the unconscious, 106–10; on the uncanny, 95, 116; on unconscious communication, 140; on unconscious wishes, 89, 95–96; on wish-fulfillment, 88–89, 91–92, 95; on word-presentations, 104–5, 106, 140
friendship, 156–57

Gallant, Jack, 7, 23–25, 78, 132, 142
gap: between intention and act, 25, 43, 91–92; of intersubjectivity, 125–26, 131–32, 152, 159, 167–68; between mind and world, 9, 10, 43, 46–47, 70–71, 85–86, 121, 122–24, 166
Gazzaniga, Michael, 104
genius, 26–28, 61, 124
givenness, 48–49, 50–51, 53, 54, 55, 60, 61, 64–65, 81; of the other, 147, 149
God, relationship to man, 168; in Derrida, 165–66; in dogmatic metaphysics, 48; in Freud, 45; in Heidegger, 51, 59–60, 66–67, 69; in Kant, 10, 45–46, 47–48, 51–52, 56–57, 60, 74–75, 118–19
Google, 90, 94, 98
Google Glass, 90
"great outdoors" (Meillassoux), 80, 168

HandShake, 8, 97
Hansen, Mark, 97–99, 102, 138, 183
Hayles, Katherine, 36
Hegel, Georg Wilhelm Friedrich, 28–29, 37, 52, 66, 82, 171n10
Hegelianism, 83, 113, 122
Heidegger, Martin: on death, 158–64, 190n81; on freedom as unfreedom, 69–71, 77–78; on God, 51, 69; on imagination, 59–60, 61, 63, 65–68; on the insecurity of mastery, 78; on intellectual intuition, 51, 59–60, 66–67, 69, 78–79, 86, 176n13; and Meillassoux, 79, 82, 86; on ontic and ontological creativity, 67; on ontological knowledge, 62, 63, 64; on the philosophy of finitude, 67, 82; on possibility and actuality, 58; on the primacy of receptivity, 50, 53, 64–68; on projecting toward possibilities, 64, 65, 68–70, 92, 158; on scientific knowledge, 82; on "standing reserve," 99; on thrownness, 67–69, 82; on time, 65, 66, 109; on the transcendental, 64–65, 67, 68–69, 70–71; on world, 68–69, 79, 82
herd, the (Nietzsche), 139–41, 165
Heroes, 135, 151
Herz, Marcus. *See* Kant, Immanuel
holism, 66, 112
horizon, projection of, 58, 63–64, 65, 67, 71, 82
human enhancement, 2, 45, 92–93, 169n4
Husserl, Edmund, 59, 79, 129, 136, 139, 143, 145–50

I and other, 12, 126, 137, 147–48
idealism, 10, 11, 29, 49, 50, 121, 124
ideas, 10–11, 13, 123–24; archetypal, 113–14; automatic execution of, 20, 22–23, 24, 43; and expression, 13–15, 25, 27–29, 30, 33, 35–37, 42; of reason, 72–73
imagination. *See* Heidegger, Martin; Kant, Immanuel
immediacy, 2, 10–11, 92, 118, 152, 166
immortality, 45, 190–91n83
implicit and explicit cognition, 38–39, 41, 137

Implicit Association Test (IAT), 99–100
improvisation, 33–34, 37, 38–43
incommunicability, 141–42
indolence, 91, 172n12. *See also* inertia
inertia, 92
Infinite Jest, 91
innermost wish, 96, 102–3
inspiration, 34, 40–41, 123–25
instrument without instrumentality, 9, 14, 23, 41
intellectual intuition. *See* Heidegger, Martin; Kant, Immanuel
intention and act, 2, 3–4, 14–15, 26, 37, 70, 78, 89–94, 129, 136
interiority, subjective, 2, 9, 79–80, 101, 125, 137–41, 145, 147, 165–68. *See also* ownness; selfhood
internal-external relation: in art, 25, 43; as condition of freedom, 46–47; in Croce, 11, 29, 32–37, 43; in Freud, 11–12, 87–89, 94–95, 107–8; in Jung, 12, 111, 117, 120–23, 166–68; in Kant, 47, 49–51, 55–58; as not applying to God, 51; as possibility and actuality, 56, 58
intersubjectivity, 125–26, 131–32, 133, 164, 167, 188n47; Derrida on, 148–49, 157; in Husserl, 145–46, 148–50; in Levinas, 147–48
intimate self-experience, 101–2, 138–39, 141–44, 182n52
intuition. *See* Croce, Benedetto; Kant, Immanuel
ipseity, 2, 160, 161. *See also* ownness; selfhood

Joyce, James, 18
Jung, Carl Gustav: on "absolute knowledge," 119–20; on archetypes of the collective unconscious, 111–12, 113–15, 117, 119–20, 121; on causality, 114, 115–19; and Hegel, 113, 122; on interiority and exteriority, 10, 11–12, 111, 117, 120–23, 166–68; and Kant, 111, 113–14, 117, 118–22; and Meillassoux, 121–22; on nonsensible knowledge, 118–121; on physics, 114–15; psychoid ontology, 115, 121–22, 125, 184n91; on the self, 112–13; on synchronicity, 10, 115–25, 166–67; *unus mundus,* 115, 119–25, 159, 166–68

Kahn, Douglas, 4
Kaku, Michio, 24
Kant, Immanuel, 14, 113–14, 117, 121, 122, 164; on aesthetic ideas, 27–28, 61; on artistic creativity, 25–28, 31, 61; on the boundaries of cognition, 48, 54–55, 56; on the consistency of experience, 59, 61, 71, 105–6, 118; on creation from nothing, 60–61, 78–79, 105–6, 118; on dogmatic/precritical metaphysics, 45, 49, 52, 57, 71; on freedom as finite, 46–47, 73–78; on free will, 71–73, 134–35, 158; on genius, 26–28, 61, 124; on God, 10, 45–46, 47–48, 51–52, 53, 56–57, 60, 74–75, 118–19; and Hegel, 28–29, 52; on holiness of the will, 74–75; on imagination, 58–63, 110; on intellectual intuition, 9, 10, 11, 47, 48–51, 54, 57, 76, 84, 118–19, 121, 134–35, 165, 168; on internal and external, 47, 49–51, 55–58; letter to Marcus Herz, 47, 49, 84, 118–19, 122; and Meillassoux, 79–81, 82, 84, 86; on the moral law, 73–77, 134, 158; on noumena, 53–54, 55, 57, 71–72, 73, 75, 76–77, 84; on the ontological proof of God, 55–56, 86; on other minds, 133–34; on overcoming finitude, 11, 75–78; on possibility and actuality, 55–57, 60; on the primacy of finitude, 52, 57; on receptivity, 9, 48–50, 60–61, 72, 78, 84, 110, 118–19; refutation of idealism, 49–50; on the schematism, 61–63, 65; on spontaneity, 50, 53, 58, 59, 72, 75; on the thing in itself, 51, 57, 71–72, 134, 177n37; on time, 17–18, 59, 61, 63, 65, 71, 109; and

transcendental apperception, 49, 54, 63, 72, 75–76; on the transcendental object, 54–55, 64
Kierkegaard, Søren, 136, 138, 158, 159
know-how, know-that, 26, 137
Kurzweil, Ray, 1, 2

Lacan, Jacques, 75, 96, 101
language: bypassing of, 131–32, 152; and consciousness, 139–41, 187n36; failure of, 141–42; and selfhood, 136–37, 138, 141, 158, 165; and the unconscious, 104–6, 140
Levinas, Emmanuel: critique of Husserl, 146–47; on the death of the other, 160–61, 163–64; on the face-to-face, 146–49, 156; on indolence, 172n12; on responsibility, 158, 159–60, 161; on the unknowability of the other, 128, 134, 135
Levy, Heinrich, 66
LeWitt, Sol, 13, 32–33
limit, passage to the, 1–2, 9, 17–18, 82, 92–93, 145, 149, 152, 167–68
London Fieldworks, 39–41
love, 100–101, 141–43, 144–45, 156
Lucier, Alvin, 20, 40, 41–42
Lyotard, Jean-François, 124

Malabou, Catherine, 8–9
materialism, 86, 101, 182n48
McLuhan, Marshall, 9, 37–38
mediation, 9–10, 33, 149, 150–52
medium, 11, 14, 23, 35, 36–38
medium specificity, 38
Meillassoux, Quentin: on the absolute, 82–83; on ancestrality, 80, 81, 83; Brassier critique of, 84–85, 86; on correlationism, 79–81, 82, 122; on facticity, 82–84, 85, 122; on God, 168; on intellectual intuition, 11, 84–86, 121–22; Toscano critique of, 85–86
mental privacy, 2, 137–38, 159
metaphysics, 48, 52, 57, 63, 71, 121–22, 162, 163, 166
Metzger, Gustav, 39–40
Metzinger, Thomas, 10, 70
MindRDR, 90
MindWave Mobile headset, 5, 90, 97
mineness. *See* ownness
Miranda, Eduardo Reck, 21–22, 23
modernism, 10–11, 18
monism, 12, 166, 184n91
moral law, the, 73–75, 76–77, 134, 158
motor imagery, 5–6, 129
Mozart, Wolfgang Amadeus, 33, 123
mundus intelligibilis, 53, 55

neo-Kantianism, 64
neurogaming, 4
neuroimaging, 3
NeuroSky, 5, 90
neurostimulation, 3, 127, 128–30
neurotechnology: and art, 10–11, 14, 24–26, 28, 31–32, 33, 37–38, 40, 41–43; definition of, 2–3; difference from standard technology, 9, 14, 23; and intellectual intuition, 45–46, 47, 48, 56–57, 60, 70, 78–79, 123; military applications of, 8, 97–98; as passage to the limit, 2, 9–10, 11, 14, 43, 89–91, 93–94, 126; politics of, 8–9. *See also* brain-computer interface; cognitive imaging
neutral language, 121, 122
new media. *See* digital media
Nietzsche, Friedrich, 124, 138–41, 153–54, 155, 159, 165
Noë, Alva, 51
notation, musical, 18–19
noumenon. *See* Kant, Immanuel
Nyman, Michael, 18–19

objectively subjective. *See* Žižek, Slavoj
oceanic, the. *See* Freud, Sigmund
originality, 61

original limitation (Schelling), 2, 9, 145
other, the: encounter with, 146–51, 156–57; responsibility for, 159–61; unknowability of, 132–36, 138, 141, 145, 151–52, 159, 164–68. *See also* I and other
other minds, problem of, 125, 133
otherness: of the other, 133, 136, 145–49, 165–66; within ourselves, 103, 139. *See also* other, the
Overgaard, Søren, 150, 188n47
ownness: as grounded in death, 158–61; in Husserl and Levinas, 146–50; as originally contaminated, 162–64, 166–68. *See also* interiority, subjective; ipseity; selfhood

panpsychism, 121
Pauli, Wolfgang, 115, 121
performativity, 13, 33–34, 153
phosphenes, 130
possibility: and actuality 2, 55–58, 74, 91–92, 94–95; as impossibility, 162–63; projection of, 69–70, 158
posttraumatic stress disorder (PTSD), 108–9
process music, 18
prostheticity, 2, 9, 45, 50, 93–94, 97
Proust, Marcel, 30, 156, 190n71
P300 response. *See* event-related potentials
pure expression, 10–11, 33–35, 41–42
pure pleasure, 89, 91–92

qualia, 143–44

receptivity. *See* Heidegger, Martin; Kant, Immanuel
resistance of reality, 70–71, 89, 92
responsibility, 73, 124, 134–35, 157–61
Roden, David, 126, 137–38, 141
Rolland, Romain, 87
Royle, Nicholas, 185n1, 190n70, 190n74

Sartre, Jean-Paul, 70, 92, 128, 149, 151, 176n14
Schelling, F. W. J., 2, 46, 66, 69, 145, 176n11
science fiction, 60–61, 102, 131
scientism, 100–101
Scott, Raymond, 20, 28, 37
secret self, 103, 138, 165–66, 168
selfhood, 2, 112, 136, 138, 149, 158, 159–61, 164, 165–66. *See also* interiority, subjective; ipseity; ownness
sender-receiver relationship: in BBI, 128–29, 131; in Derrida, 12, 152–57, 166–67
skill, 13, 14, 22, 23, 25–26, 31, 32, 41–42
smart homes, 90–91
Solaris, 102, 105
solipsism, 146, 150, 164
Spinoza, Benedict de, 184n91, 191n83
spontaneity, 71, 134–35, 158, 160; in art, 33–34, 38; of thought, 50, 53, 59, 61, 65–66, 67–68, 72, 75
Stalker, 96, 102–3
Stelarc, 42
Stiegler, Bernard, 8–9, 50, 94
Stockhausen, Karlheinz, 18, 19–20, 22, 38–39, 40, 41, 42
Swedenborg, Emanuel, 115–117
synchronicity. *See* Jung, Carl Gustav

Tarkovsky, Andrei, 96, 102, 181n34
technological advancement, 1–2, 12, 37–38, 45, 137; in Freud, 89, 92, 95, 140
technological enhancement. *See* human enhancement
telecommunication, 90, 131, 132, 150, 186n16
teleiopoesis. *See* Derrida, Jacques
telepathy, 10; in Derrida, 150–52, 153, 155, 157, 166–68; in Freud, 140; and the other, 12, 126, 132, 133, 141, 145, 150, 159, 164; in popular culture, 127–28, 135; and science, 127; and synchronicity, 166–68
Toscano, Alberto, 85–86

Transcranial Magnetic Stimulation (TMS), 128–29
transhumanism, 169n4

unconscious bias, 99–100
universality and singularity, 28, 111, 117–18, 136–41, 159, 164–68
unus mundus. See Jung, Carl Gustav

Vidal, Jacques, 4–5
violence, 146, 147–49
Virilio, Paul, 93

Walter, W. Grey, 4–6
why finitude, question of, 52, 67, 81, 145
Wilson, Brian, 33
wish-fulfillment: in Freud, 11, 88–89, 91–92, 95, 96; as nightmare, 95, 100, 102; in Sartre, 70
Wittgenstein, Ludwig, 136–37, 143, 151, 188n47

Žižek, Slavoj: on freedom as trans-phenomenal, 75–77; on fundamental fantasy, 100–101; on the id-machine, 102–3, 105; on neurotechnology, 9, 11, 43, 45–46, 47, 71, 78, 93, 122–23; on new media, 37; on the objectively subjective, 100; on other minds, 133, 149
Zuckerberg, Mark, 128

(continued from page ii)

23 Vampyroteuthis Infernalis: A Treatise, with a Report by the Institut Scientifique de Recherche Paranaturaliste
Vilém Flusser and Louis Bec

22 Body Drift: Butler, Hayles, Haraway
Arthur Kroker

21 HumAnimal: Race, Law, Language
Kalpana Rahita Seshadri

20 Alien Phenomenology, or What It's Like to Be a Thing
Ian Bogost

19 CIFERAE: A Bestiary in Five Fingers
Tom Tyler

18 Improper Life: Technology and Biopolitics from Heidegger to Agamben
Timothy C. Campbell

17 Surface Encounters: Thinking with Animals and Art
Ron Broglio

16 Against Ecological Sovereignty: Ethics, Biopolitics, and Saving the Natural World
Mick Smith

15 Animal Stories: Narrating across Species Lines
Susan McHugh

14 Human Error: Species-Being and Media Machines
Dominic Pettman

13 Junkware
Thierry Bardini

12 A Foray into the Worlds of Animals and Humans, *with* A Theory of Meaning
Jakob von Uexküll

11 Insect Media: An Archaeology of Animals and Technology
Jussi Parikka

10 Cosmopolitics II
Isabelle Stengers

9 Cosmopolitics I
Isabelle Stengers

8 What Is Posthumanism?
Cary Wolfe

7 Political Affect: Connecting the Social and the Somatic
John Protevi

6 Animal Capital: Rendering Life in Biopolitical Times
Nicole Shukin

5 Dorsality: Thinking Back through Technology and Politics
David Wills

4 Bíos: Biopolitics and Philosophy
Roberto Esposito

3 When Species Meet
Donna J. Haraway

2 The Poetics of DNA
Judith Roof

1 The Parasite
Michel Serres

MICHAEL HAWORTH completed his PhD at Goldsmiths College, University of London, where he taught history of art for several years. He has published widely on philosophical aesthetics, technology, and phenomenology.